T0182086

Artificial Intelligence in Medical Sciences and Psychology

With Application of Machine Language, Computer Vision, and NLP Techniques

Tshepo Chris Nokeri

Apress®

Artificial Intelligence in Medical Sciences and Psychology: With Application of Machine Language, Computer Vision, and NLP Techniques

Tshepo Chris Nokeri
Pretoria, South Africa

ISBN-13 (pbk): 978-1-4842-8216-8 ISBN-13 (electronic): 978-1-4842-8217-5
https://doi.org/10.1007/978-1-4842-8217-5

Managing Director, Apress Media LLC: Welmoed Spahr
Acquisitions Editor: Celestin Suresh John
Development Editor: Laura Berendson
Coordinating Editor: Mark Powers
Copy Editor: Mary Behr

Cover designed by eStudioCalamar

Cover image by Jason Leung on Unsplash (www.unsplash.com)

Distributed to the book trade worldwide by Apress Media, LLC, 1 New York Plaza, New York, NY 10004, U.S.A. Phone 1-800-SPRINGER, fax (201) 348-4505, e-mail orders-ny@springer-sbm.com, or visit www.springeronline.com. Apress Media, LLC is a California LLC and the sole member (owner) is Springer Science + Business Media Finance Inc (SSBM Finance Inc). SSBM Finance Inc is a **Delaware** corporation.

For information on translations, please e-mail booktranslations@springernature.com; for reprint, paperback, or audio rights, please e-mail bookpermissions@springernature.com.

Apress titles may be purchased in bulk for academic, corporate, or promotional use. eBook versions and licenses are also available for most titles. For more information, reference our Print and eBook Bulk Sales web page at www.apress.com/bulk-sales.

Any source code or other supplementary material referenced by the author in this book is available to readers on GitHub (https://github.com/Apress). For more detailed information, please visit www.apress.com/source-code.

Printed on acid-free paper

Table of Contents

About the Author

Tshepo Chris Nokeri harnesses advanced analytics and artificial intelligence to foster innovation and optimize business performance. He delivers complex solutions to companies in the mining, petroleum, and manufacturing industries. He received a bachelor's degree in information management. He graduated with honors in business science from the University of the Witwatersrand, Johannesburg, on a Tata Prestigious Scholarship and a Wits Postgraduate Merit Award. He was unanimously awarded the Oxford University Press Prize. Tshepo has authored four books: *Data Science Revealed* (Apress, 2021), *Implementing Machine Learning in Finance* (Apress, 2021), *Web App Development and Real-Time Web Analytics with Python* (Apress, 2021), and *Econometrics and Data Science* (Apress, 2022).

About the Technical Reviewer

Ashish Soni is an experienced AIML consultant and solutions architect. He has worked and solved business problems related to computer vision, natural language processing, machine learning, artificial intelligence, data science, statistical analysis, data mining, and cloud computing. Ashish holds a B. Tech. degree in Chemical Engineering from Indian Institute of Technology, Bombay, India; a master's degree in Economics; and a post graduate diploma in Applied Statistics. He has worked across different industry areas such as finance, healthcare, education, sports, human resources, retail, and logistics automation. He currently works with a technology services company based out of Bangalore.

CHAPTER 1

An Introduction to Artificial Intelligence in Medical Sciences and Psychology

In this chapter, I'll establish the specific context and structure of the book, and then offer an overview of the varying medical specialties central to it before covering independent subsets of artificial intelligence. I then run through some valuable tools for undertaking exercises, such as the Python programming language, distribution package, and libraries. Lastly, I'll acquaint you with different algorithms, including when to carry them out.

Disclaimer This book is not a medical article or textbook; rather, it is a programming technical book that demonstrates approaches of properly implementing subsets of artificial intelligence to discover patterns in health sciences. None of the content in this book is health advice.

T. C. Nokeri, *Artificial Intelligence in Medical Sciences and Psychology*,
https://doi.org/10.1007/978-1-4842-8217-5_1

Context of the Book

This book is not a medical science textbook, nor am I a medical scientist.

The aim of this book is to offer health science practitioners a way of implementing machine learning to simplify their practices. The book does not cover key theoretical concepts underpinning medical fields, as this is outside the scope of this book.

The Book's Central Point

Figure 1-1 shows the medical specialties central to this book.

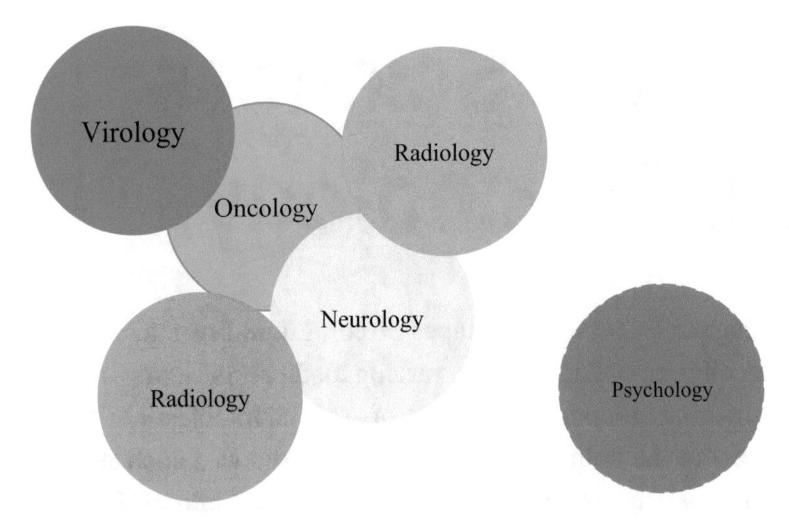

Figure 1-1. *Medical specialities central to this book*

I have centered this book on medical and mental health sciences. I present an approach for implementing machine learning to tackle data that is related to medical sciences (i.e., oncology, neuroscience, and cardiology) and social science (i.e., psychology).

Artificial Intelligence Subsets Covered in this Book

Figure 1-2 exhibits the subsets of artificial intelligence that this book covers to help you understand the practical implications of artificial intelligence in medical sciences.

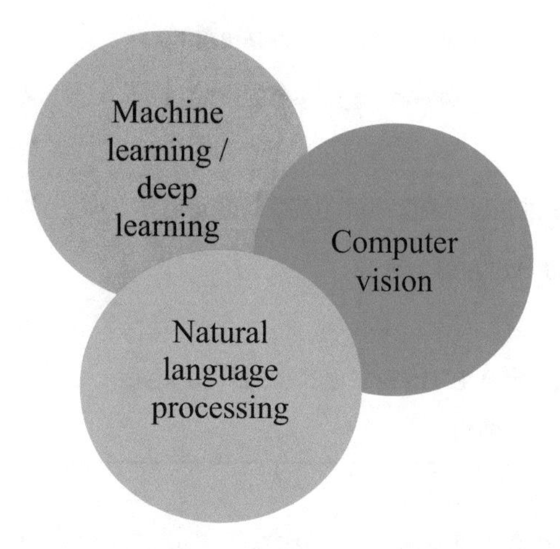

***Figure 1-2.** Artificial intelligence subsets covered in this book*

This book covers three subsets of artificial intelligence (i.e., machine learning, computer vision, and natural language).

Structure of the Book

Table 1-1 outlines the structure of the book.

Table 1-1. *The Structure of This Book*

Chapter	Subset	Medical specialty
Chapter 2	Deep learning	Cardiology
Chapter 3	Markov's method and Monte Carlo simulation method	Virology
Chapter 4	Computer vision and deep learning	Oncology
Chapter 5	Computer vision and deep learning	Neurology and radiology
Chapter 6	Computer vision and deep learning	Virology
Chapter 7	Survival analysis	Oncology (clinical trials)
Chapter 8	Machine learning and natural language processing	General

This book covers machine learning, computer vision, and natural language processing. In addition, it also covers survival regression analysis, Hidden Markov decision-making, and Monte Carlo simulation. It includes key medical specialties (i.e., cardiology, oncology, and neurology, among others).

Before you proceed with this book, make sure you understand key concepts in statistics and key health sciences. Although unnecessary, besides proper comprehension of relevant concepts, some background experience in programming using the Python programming language will help.

Tools Used in This Book

This book implements the Python programming language to perform typical research tasks that are associated with investigating patterns in medical-related data. Python is an open-source programming language that is relatively easy to learn. It supports us when programming applications and implementing scientific research tasks via a human-like language, therefore gaining the skills for programming will not be difficult. It is a suitable candidate for implementing and scaling artificial intelligence projects.

Python Distribution Package

To make your Python programming experience easier, you will employ a distribution package that efficiently manages key Python resources and functional dependencies. There are myriad Python distribution packages (i.e., Anaconda, etc.) and integrated development environments (i.e., PyCharm).

Anaconda Distribution Package

Anaconda is the most prevalent Python distribution package. In addition, it supports other programming languages (i.e., R, Scala, PySpark, etc.). The comprehensive package contains environments like JupyterLab, Jupyter Notebook, and Spyder. Plus, it holds a cmd and a PowerShell terminal to execute scripts.

Jupyter Notebook

Jupyter Notebook is the most prevalent interactive environment for writing Python programming. It enables you to view the output of your code inline. You do not have to write, assemble, or test the entire program. It is a suitable candidate for prototyping.

Python Libraries

The availability of a global pool of open-source libraries sets the Python programming language apart from other languages. A library is an elaborate set of prewritten programming code that allows you to perform complex programming tasks with little effort. There is a large pool of standard open-source Python libraries.

Table 1-2 outlines the libraries that this book implements, including their basic usage and installation.

Table 1-2. *Primary Python Libraries Used in This Book*

Name	Usage	Installation
Matplotlib	2D static charting	For a Python environment, use `pip install matplotlib`. For a conda environment, use `conda install -c conda-forge matplotlib`.
seaborn	2D static charting	For a Python environment, use `pip install seaborn`. For a conda environment, use `conda install -c anaconda seaborn`.
pandas	Static tabulation and data manipulation	For a Python environment, use `pip install pandas`. For a conda environment, use `conda install -c anaconda pandas`.
NumPy	Numeric computation and data manipulation	For a Python environment, use `pip install numpy`. For a conda environment, use `conda install -c anaconda numpy`.

(*continued*)

Table 1-2. (*continued*)

Name	Usage	Installation
CV2	Computer vision and image data manipulation	For a Python environment, use `pip install opencv-python`. For a conda environment, use `conda install -c conda-forge opencv`.
TensorFlow	Neural network development	For a Python environment, use `pip install tensorflow`. For a conda environment, use `conda install -c conda-forge tensorflow`.
Keras	Neural network development	For a Python environment, use `pip install keras`. For a conda environment, use `conda install -c conda-forge keras`.
scikit-learn	Preprocessing, model selection, development, and evaluation	For a Python environment, use `pip install scikit-learn`. For a conda environment, use `conda install -c anaconda scikit-learn`.
Factor Analyzer	Exploratory factor analysis	For a Python environment, use `pip install factor_analyzer`. For a conda environment, use `conda install -c ets factor_analyzer`.

Encapsulating Artificial Intelligence

To recognize artificial intelligence, first you must understand human intelligence, which involves scanning information in the environment, modeling it, finding patterns to make conclusions, and then using such conclusions as a basis of action. The unified field of artificial intelligence intends to program computer systems in ways that mirror human intelligence. In simple terms, deep learning attempts to mirror the underlying process of the human brain; computer vision attempts to mirror the underlying process of the human visual system; and natural language processing attempts to mirror the approach of human process language.

Implementing Algorithms

This book will implement myriad algorithms to predict an array of instances from various datasets. An algorithm is a logical sequence of steps that involves inheriting instances of features, modeling, and computing instances of a feature with support of a function. You can consider algorithms as formulas.

Figure 1-3 exhibits key algorithms.

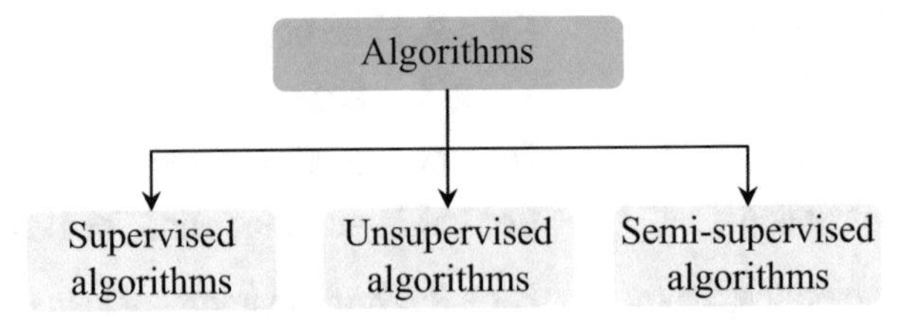

Figure 1-3. *Key algorithms*

Supervised Algorithms

Figure 1-4 exhibits the primary families of supervised algorithms.

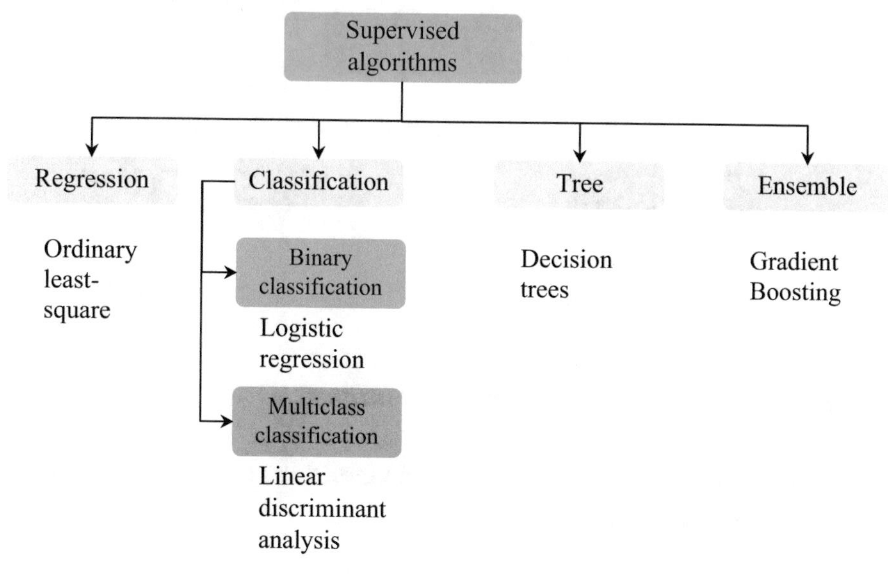

Figure 1-4. *Supervised algorithm families*

The primary families of supervised algorithms includes the regression family (i.e., ordinary least-square), the classification family (i.e., logistic regression), the tree family, and the ensemble family (i.e., random forest trees).

Unsupervised Algorithms

Figure 1-5 depicts the primary families of unsupervised algorithms.

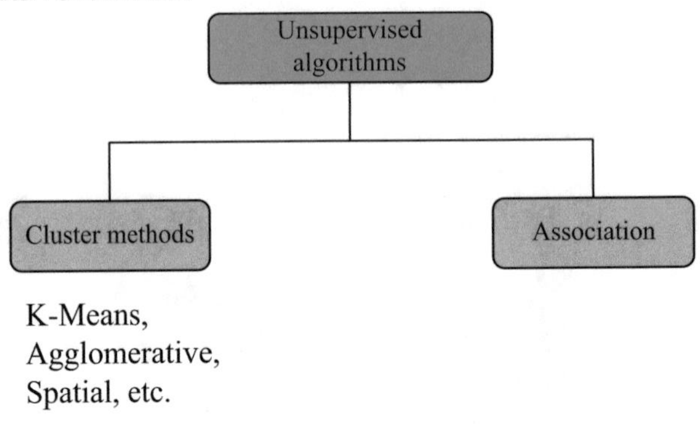

Figure 1-5. *Unsupervised algorithm families*

The primary families of unsupervised algorithms are the cluster family and association algorithms.

Artificial Neural Networks

Table 1-3 depicts the artificial neural networks that this book implements, including their usage activation function and layers.

Table 1-3. *Primary Artificial Neural Networks Employed in This Book*

Name	Usage	Activation function	Layers
Deep Belief Network	Classifying diabetes and cardiovascular disease	Relu and sigmoid	Dense and Dropout
Convolutional neural networks	Classifying MRI, cancer (breast, skin, and brain cancer), and X-ray scans	Relu and Softmax	Conv2D, MaxPooling2D, Flatten, and Dropout

The subsequent chapters will acquaint you with the different activation functions, layers, and their usage.

Conclusion

Now that you have a sense of the material I'll be covering in this book and the tools you will need to work your way through the various exercises, let's dive right in and explore using neural networks to detect patterns in diseases.

CHAPTER 2

Realizing Patterns in Diseases with Neural Networks

This chapter covers the application of artificial neural networks in modeling medical data. You'll learn how to execute deep belief networks to model data and predict whether a patient suffers from a disease (specifically cardiovascular disease and diabetes). Further, you will learn how to appraise networks with key metrics to find out the extent to which the networks differentiate patients who suffer from the disease from those who do not.

Classifying Cardiovascular Disease Diagnosis Outcome Data by Executing a Deep Belief Network

In this section, you will learn how to classify patients with cardiovascular diseases by executing a deep belief network. Cardiovascular disease signifies a disorder in the heart and blood vessels. You may download the dataset from Kaggle[1].

[1] www.kaggle.com/sulianova/cardiovascular-disease-dataset

T. C. Nokeri, *Artificial Intelligence in Medical Sciences and Psychology*,
https://doi.org/10.1007/978-1-4842-8217-5_2

Listing 2-1 collects data relating to cardiovascular disease from the CSV file and removes the column titled id. Start by installing pandas in your environment: pip install pandas.

Listing 2-1. Collecting Data

```
import pandas as pd
cardiovascular_data = pd.read_csv(r"filepath\cardio_train.csv",
sep=";")
cardiovascular_data.drop(["id"], axis = 1, inplace = True)
```

Listing 2-2 processes the column titled age so that it is suitable for analysis.

Listing 2-2. Age Rounding

```
cardiovascular_data["age"] = round(cardiovascular_data["age"] /
365.25, 2)
```

Listing 2-3 constructs a pair plot, which exhibits how features relate to each other in the cardiovascular dataset. In addition, it depicts the distribution of the features (see Figure 2-1). Start by installing Matplotlib in your environment: pip install matplotlib.

Listing 2-3. Constructing a Pair Plot

```
import matplotlib.pyplot as plt
%matplotlib inline
import seaborn as sns
sns.set("talk","ticks", font_scale = 1, font = "Calibri")
sns.pairplot(cardiovascular_data)
plt.show()
```

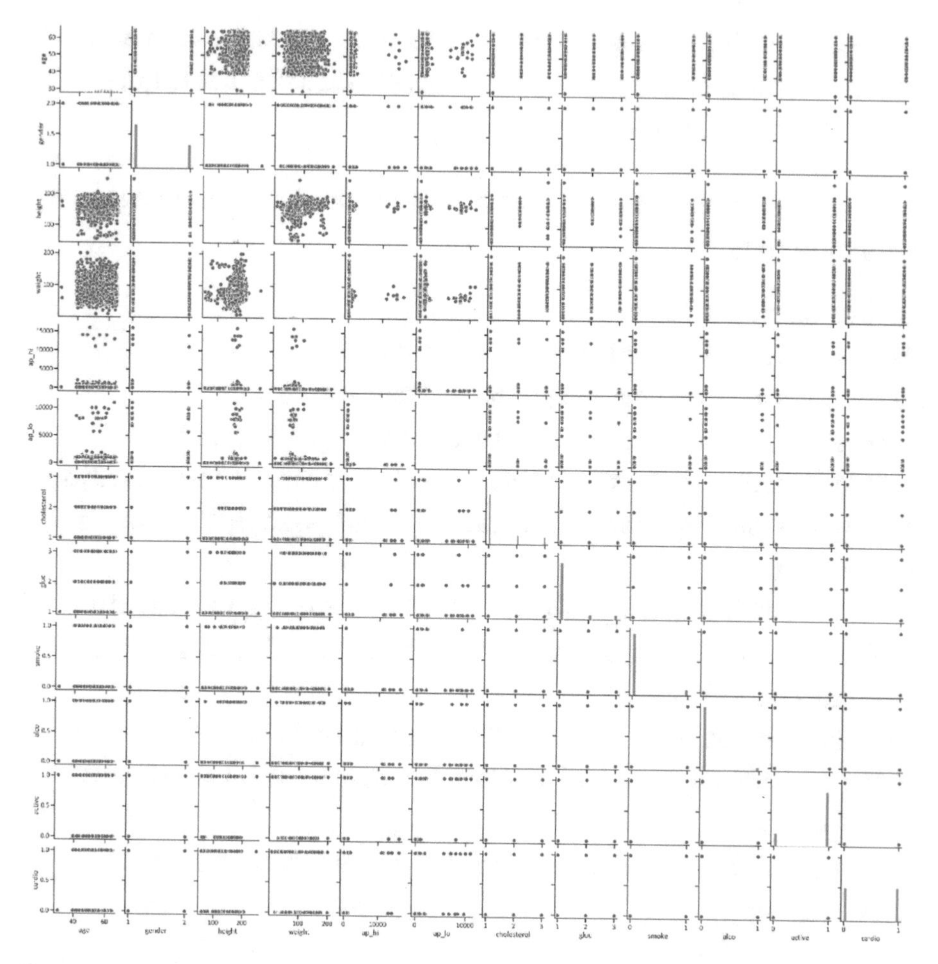

Figure 2-1. *Pair plot*

Preprocessing the Cardiovascular Disease Diagnosis Outcome Data

Listing 2-4 preprocesses the cardiovascular disease diagnosis data. To begin, it imports crucial libraries for preprocessing. Subsequently, it places features in NumPy arrays. Then, it separates the values of features for training and testing the artificial neural network. It concludes by standardizing the training-independent features. Start by installing NumPy in your environment: `pip install numpy`. Also install scikit-learn: `pip install -U scikit-learn`.

Listing 2-4. Preprocessing the Cardiovascular Diseases Diagnosis Outcome Data

```
import numpy as np
from sklearn.model_selection import train_test_split
from sklearn.preprocessing import StandardScaler
from sklearn.model_selection import train_test_split
x_train_cardio, x_test_cardio, y_train_cardio, y_test_cardio =
train_test_split(x_cardio, y_cardio, test_size = 0.2, random_
state = 0)
x_train_cardio, x_val_cardio, y_train_cardio, y_val_cardio =
train_test_split(x_train_cardio, y_train_cardio, test_size =
0.1, random_state = 0)
standard_scaler_for_cardio = StandardScaler()
x_train_cardio = standard_scaler_for_cardio.fit_transform(x_
train_cardio)
x_test_cardio = standard_scaler_for_cardio.transform(x_
test_cardio)
```

Debunking Deep Belief Networks

A deep belief network lumps together manifold restricted Boltzmann machines by permitting innumerable hidden layers, thus magnifying the complexity of a neural network. Equation 2-1 defines a deep belief network:

$$E(w)=\sum_{i=1}^{n} e\left(f\left(x_i, w\right), y_i\right). \qquad \text{(Equation 2-1)}$$

Designing the Deep Belief Network

Listing 2-5 develops the artificial neural network. To begin, it imports crucial libraries for developing the network. Following that, it outlines the structure of the network in such a way that the input layer holds 11 neurons with a `relu` activation, three hidden layers hold 11 neurons with a `relu` activation function, and the output layer generates labels with a sigmoid function (see Figure 2-2). Start by installing TensorFlow in your environment: pip install `tensorflow`.

Listing 2-5. Structuring the Deep Belief Network

```
import tensorflow as tf
from tensorflow.keras import Sequential, regularizers
from tensorflow.keras.layers import Dense
def cardiovascular_dbn_function():
    cardiovascular_dbn_model = Sequential()
    cardiovascular_dbn_model.add(Dense(11, input_dim = 11,
    activation = "relu"))
    cardiovascular_dbn_model.add(Dense(11, activation = "relu"))
    cardiovascular_dbn_model.add(Dense(11, activation = "relu"))
    cardiovascular_dbn_model.add(Dense(11, activation = "relu"))
    cardiovascular_dbn_model.add(Dense(1, activation =
    "sigmoid"))
    cardiovascular_dbn_model.compile(loss = "binary_
    crossentropy", optimizer = "adam", metrics = ["accuracy"])
    return cardiovascular_dbn_model
```

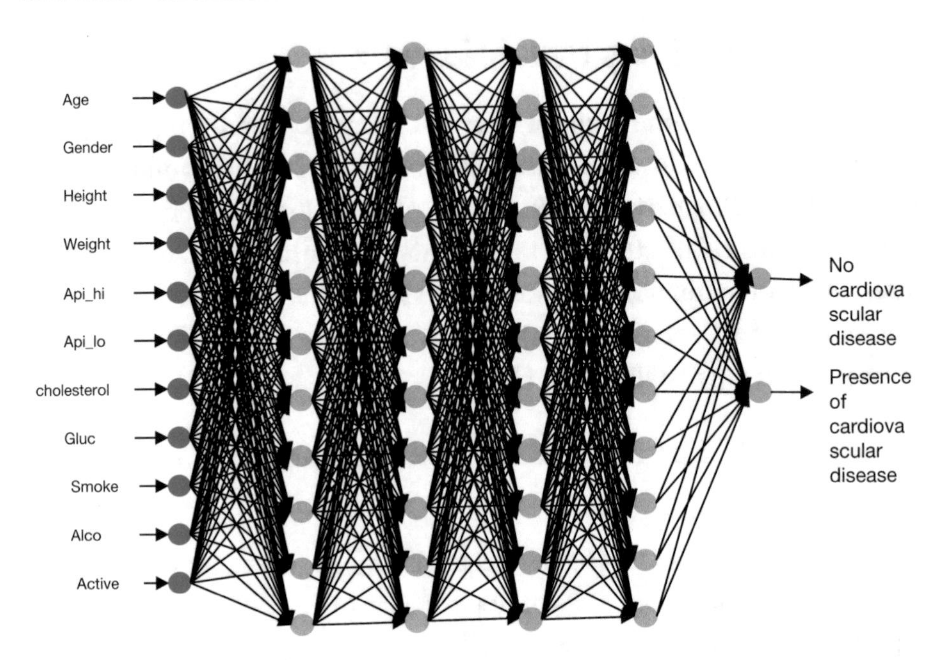

Figure 2-2. *Neural network structure*

Relu Activation Function

Figure 2-2 depicts that the deep belief network encloses the `relu` activation function in the input layer.

Equation 2-2 defines a relu activation function:

$$Relu = max(0,x). \quad \text{(Equation 2-2)}$$

Equation 2-2 suggests that the independent feature holds any value from 0.

Sigmoid Activation Function

Equally, Equation 2-3 defines a sigmoid activation in the output layer:

$$Sigmoid = \frac{1}{1+e^{-x}}. \qquad \text{(Equation 2-3)}$$

Equation 2-3 generates classes (either 0 or 1).

Listing 2-6 encloses the deep belief network.

Listing 2-6. Enclosing the Deep Belief Network

```
from keras.wrappers.scikit_learn import KerasClassifier
cardiovascular_dbn_model = KerasClassifier(build_fn =
cardiovascular_dbn_function)
```

Training the Deep Belief Network

Listing 2-7 trains the artificial neural network on the data. Equally, it outlines the validation data, including the number of epochs (equally recognized as iterations) and batch size (equally recognized as the number of samples the artificial neural network learns at each iteration).

Listing 2-7. Training the Deep Belief Network

```
cardiovascular_dbn_model_history = cardiovascular_dbn_model.
fit(x_train_cardio, y_train_cardio, validation_data = (x_val_
cardio, y_val_cardio), epochs = 64, batch_size = 16)
cardiovascular_dbn_model_history
```

Outlining the Deep Belief Network's Predictions

Listing 2-8 outlines the deep belief network's predictions (see Table 2-1).

Listing 2-8. Outlining the Deep Belief Network's Predictions

```
y_hat_cardiovascular_dbn_model = cardiovascular_dbn_model.
predict(x_test_cardio)
actual_cardio = pd.DataFrame(y_test_cardio)
actual_cardio.columns = ["Actual"]
predicted_cardio = pd.DataFrame(y_test_cardio)
predicted_cardio.columns = ["Predicted"]
actual_and_predicted_cardio = pd.concat([actual_cardio,
predicted_cardio], axis = 1)
actual_and_predicted_cardio.loc[actual_and_predicted_cardio.
Actual == 0, "Actual"] = "No cardiovascular disease"
actual_and_predicted_cardio.loc[actual_and_predicted_cardio.
Actual == 1, "Actual"] = "Presence of cardiovascular disease"
actual_and_predicted_cardio.loc[actual_and_predicted_cardio.
Predicted == 0, "Predicted"] = "No cardiovascular disease"
actual_and_predicted_cardio.loc[actual_and_predicted_cardio.
Predicted == 1, "Predicted"] = "Presence of cardiovascular
disease"
actual_and_predicted_cardio
```

Table 2-1. *The Deep Belief Network's Predictions*

	Actual	Predicted
0	No cardiovascular disease	No cardiovascular disease
1	No cardiovascular disease	No cardiovascular disease
2	No cardiovascular disease	No cardiovascular disease
3	No cardiovascular disease	No cardiovascular disease
4	No cardiovascular disease	No cardiovascular disease
...	...	...
13995	Presence of cardiovascular disease	Presence of cardiovascular disease
13996	No cardiovascular disease	No cardiovascular disease
13997	Presence of cardiovascular disease	Presence of cardiovascular disease
13998	No cardiovascular disease	No cardiovascular disease
13999	Presence of cardiovascular disease	Presence of cardiovascular disease

Considering the Deep Neural Network's Performance

To determine how well the deep neural network classifies patient cardiovascular disease outcomes in training and cross-validation, this segment monitors the degree of binary cross-entropy loss and accuracy metric fluctuations as epochs increase. First, let's look at the confusion matrix and then the classification report.

Listing 2-9 outlines the deep belief network's predictions (see Table 2-2).

Listing 2-9. Outlining the Deep Belief Network's Confusion Matrix

```
from sklearn.metrics import confusion_matrix
cardiovascular_dbn_model_confusion_matrix =
pd.DataFrame(confusion_matrix(y_test_cardio,
                               y_hat_cardiovascular_
                               dbn_model),
                         index = ["Actual: No cardiovascular
                         disease",
                              "Actual: Presence of
                              cardiovascular disease"],
                         columns = ("Predicted: No
                         cardiovascular disease",
                              "Predicted: Presence of
                              cardiovascular disease"))
cardiovascular_dbn_model_confusion_matrix
```

Table 2-2. *The Deep Belief Network's Confusion Matrix*

	Predicted: No cardiovascular disease	Predicted: Presence of cardiovascular disease
Actual: No cardiovascular disease	5592	1477
Actual: Presence of cardiovascular disease	2279	4652

Table 2-2 shows that the deep belief network

- Correctly predicted patients had no cardiovascular disease 5592 times.
- Incorrectly predicted that patients had cardiovascular disease when they did not have it 1477 times.

- Incorrectly predicted that patients had no cardiovascular disease when they did have it 2279 times.
- Correctly predicted patients had a cardiovascular disease 4652 times.

Listing 2-10 outlines the artificial neural network's classification report, which holds the accuracy score, precision score, recall, f-1 score, and support (see Table 2-3).

Equation 2-4 defines the precision:

$$precision = \frac{TP}{TP + FP}. \quad \text{(Equation 2-4)}$$

True positive (TP) counts when a model estimates a positive response and it is correct, and false positive (FP) counts when a model estimates a negative class.

Equation 2-5 defines the recall:

$$recall = \frac{TP}{TP + FN}. \quad \text{(Equation 2-5)}$$

False negative (FN) counts when a model estimates a positive class, but it was a negative.

Equation 2-6 defines the F1 score:

$$F1 = \frac{2 \times precision \times recall}{precision + recall}. \quad \text{(Equation 2-6)}$$

Equation 2-7 defines the accuracy score:

$$accuracy = \frac{TP + TN}{TP + FN + TN + FP}. \quad \text{(Equation 2-7)}$$

Listing 2-10. Outlining the Deep Belief Network's Classification Report

```
from sklearn.metrics import classification_report
cardiovascular_dbn_model_report = pd.DataFrame(classification_
report(y_test_cardio, y_hat_cardiovascular_dbn_model,
                                output_dict = True)).transpose()
cardiovascular_dbn_model_report
```

Table 2-3. *The Deep Belief Network's Classification Report*

	Precision	Recall	F-1 score	Support
0	0.707317	0.804074	0.752598	7069.000000
1	0.767773	0.660655	0.710198	6931.000000
accuracy	0.733071	0.733071	0.733071	0.733071
macro avg	0.737545	0.732365	0.731398	14000.000000
weighted avg	0.737247	0.733071	0.731607	14000.000000

Table 2-3 suggests that the deep belief network is precise 71% of the time when predicting the absence of cardiovascular disease and 77% precise when predicting its presence. The overall accuracy score is 73%.

Accuracy Fluctuations Across Epochs in Training and Cross-Validation

Listing 2-11 and Figure 2-3 depict the degree of accuracy fluctuations as epochs increase in training and cross-validation when the deep neural network classifies patient cardiovascular disease outcomes.

Listing 2-11. Charting Accuracy Fluctuations Across Epochs in Training and Cross-Validation

```
plt.plot(cardiovascular_dbn_model_history.history["accuracy"],
         color = "orange",
         marker = "o",
         label = "Training accuracy")
plt.plot(cardiovascular_dbn_model_history.history["val_
accuracy"],
         color = "blue",
         marker = "o",
         label = "CV accuracy")
plt.xlabel("Epochs")
plt.ylabel("Loss")
plt.legend(loc = "best")
plt.show()
```

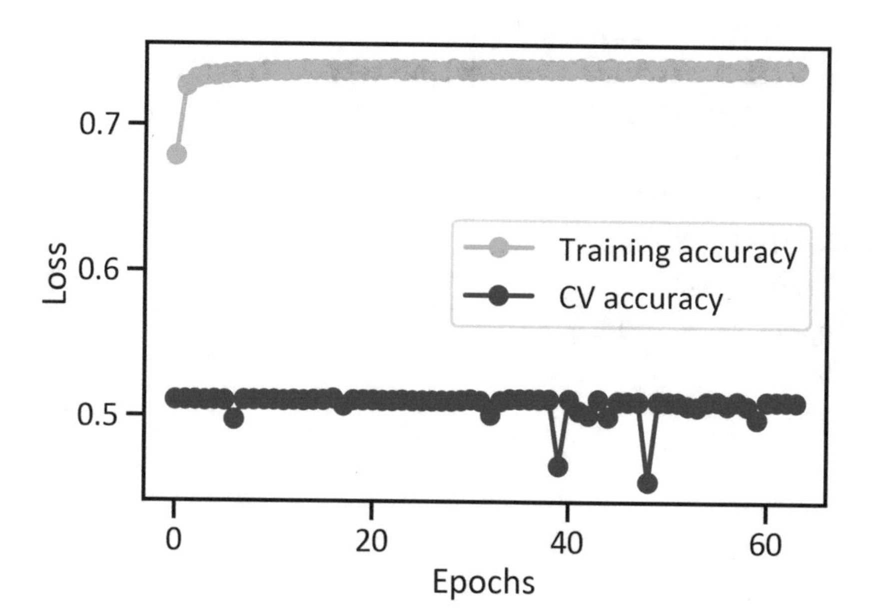

Figure 2-3. *Accuracy fluctuations across epochs in training and cross-validation*

Figure 2-3 depicts that the accuracy is higher in cross-validation than in training when classifying patient cardiovascular disease outcomes.

Binary Cross-Entropy Loss Fluctuations Across Epochs in Training and Cross-Validation

Figure 2-4 depicts the degree of binary cross-entropy loss fluctuations as epochs increase in training and cross-validation when the deep neural network classifies patient cardiovascular disease outcomes. See Listing 2-12 for the code.

Listing 2-12. Binary Cross-Entropy Loss Fluctuations Across Epochs in Training and Cross-Validation

```
plt.plot(cardiovascular_dbn_model_history.history["loss"],
         color = "orange",
         marker = "o",
         label = "Training loss")
plt.plot(cardiovascular_dbn_model_history.history["val_loss"],
         color = "blue",
         marker = "o",
         label = "CV loss")
plt.xlabel("Epochs")
plt.ylabel("Loss")
plt.legend(loc = "best")
plt.show()
```

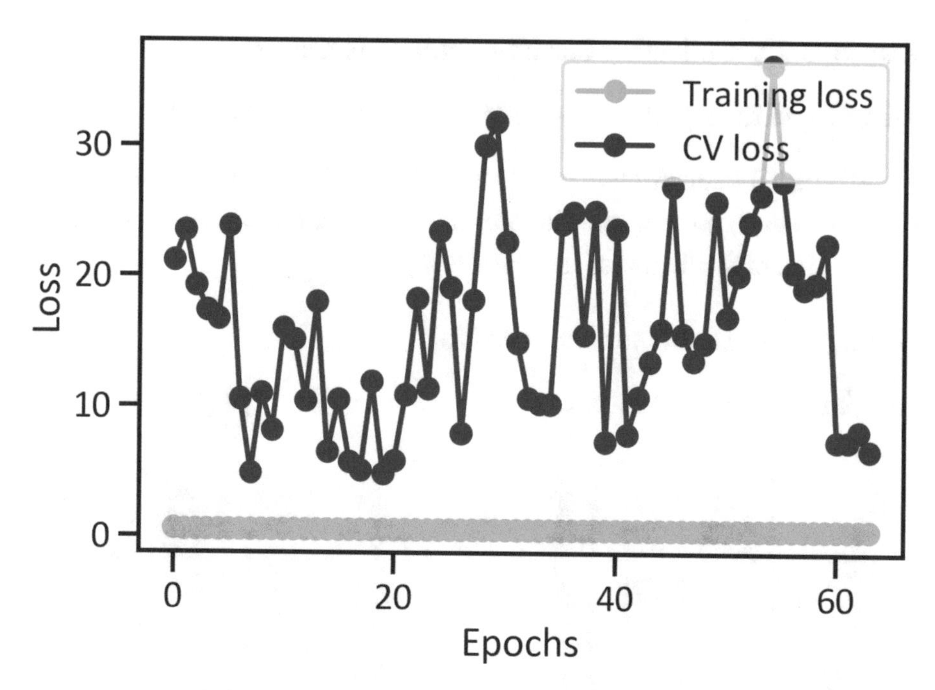

Figure 2-4. *Binary cross-entropy loss fluctuations across epochs in training and cross-validation*

Figure 2-4 shows that the binary cross-entropy loss in training fluctuates a lot in cross-validation but is constant in training when classifying patient cardiovascular disease outcomes.

Classifying Diabetes Diagnosis Outcome Data by Executing a Deep Belief Network

Listing 2-13 collects data relating to cardiovascular disease from a CSV file. You may download the dataset from Kaggle[2].

[2] www.kaggle.com/uciml/pima-indians-diabetes-database

Listing 2-13. Collecting Cardiovascular Disease

```
diabetes_data = pd.read_csv(r"filepath\diabetes.csv")
```

Listing 2-14 constructs a pair plot, which exhibits how features relate with each other in the diabetes dataset. In addition, it depicts the distribution of the features (see Figure 2-5).

Listing 2-14. Diabetes Data Pair Plot

```
sns.pairplot(diabetes_data)
```

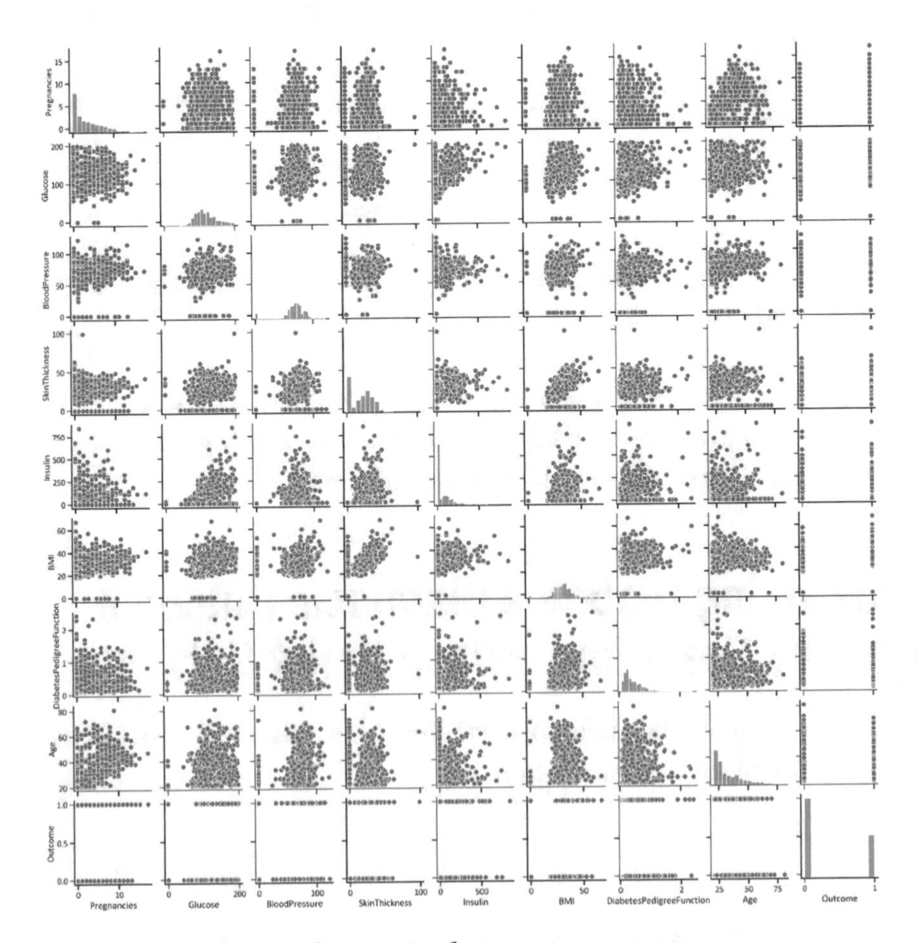

Figure 2-5. *Diabetes data pair plot*

Executing a Deep Belief Network to Classify Diabetes Diagnosis Outcome Data

Listing 2-15 allots features to NumPy arrays. Subsequently, it separates the values of features for training and testing the deep belief network. Then, it standardizes the training-independent features. It concludes by training the network on the diabetes data.

The network is structured in such a way that the input layer holds 11 neurons with a relu activation, three hidden layers hold 11 neurons with a relu activation function, and the output layer generates labels with a sigmoid function (see Figure 2-6).

Listing 2-15. Executing a Deep Belief Network to Classify Diabetes Diagnosis Outcome Data

```
x_diabetes = np.array(diabetes_data.iloc[::, 0:8])
y_diabetes = np.array(diabetes_data.iloc[::, -1])
x_train_diabetes, x_test_diabetes, y_train_diabetes, y_test_
diabetes = train_test_split(x_diabetes, y_diabetes, test_size =
0.2, random_state = 0)
x_train_diabetes, x_val_diabetes, y_train_diabetes, y_val_
diabetes = train_test_split(x_train_diabetes, y_train_diabetes,
test_size = 0.1, random_state = 0)
standard_scaler_for_diabetes = StandardScaler()
x_train_diabetes = standard_scaler_for_diabetes.fit_
transform(x_train_diabetes)
x_test_diabetes = standard_scaler_for_diabetes.transform(x_
test_diabetes)
from tensorflow.python.keras.layers import Dropout
def diabetes_dbn_function():
    diabetes_dbn_model = Sequential()
```

```
    diabetes_dbn_model.add(Dense(8, input_dim = 8,
    activation="relu"))
    diabetes_dbn_model.add(Dropout(0.2))
    diabetes_dbn_model.add(Dense(8, activation = "relu"))
    diabetes_dbn_model.add(Dense(8, activation = "relu"))
    diabetes_dbn_model.add(Dense(8, activation = "relu"))
    diabetes_dbn_model.add(Dense(1, activation = "sigmoid"))
    diabetes_dbn_model.compile(loss = "binary_crossentropy",
    optimizer = "adam", metrics = ["accuracy"])
    return diabetes_dbn_model
diabetes_dbn_model = KerasClassifier(build_fn = diabetes_dbn_
function)
diabetes_dbn_model_history = diabetes_dbn_model.fit(x_train_
diabetes, y_train_diabetes, validation_data = (x_val_diabetes,
y_val_diabetes), epochs = 64, batch_size = 16)
diabetes_dbn_model_history
```

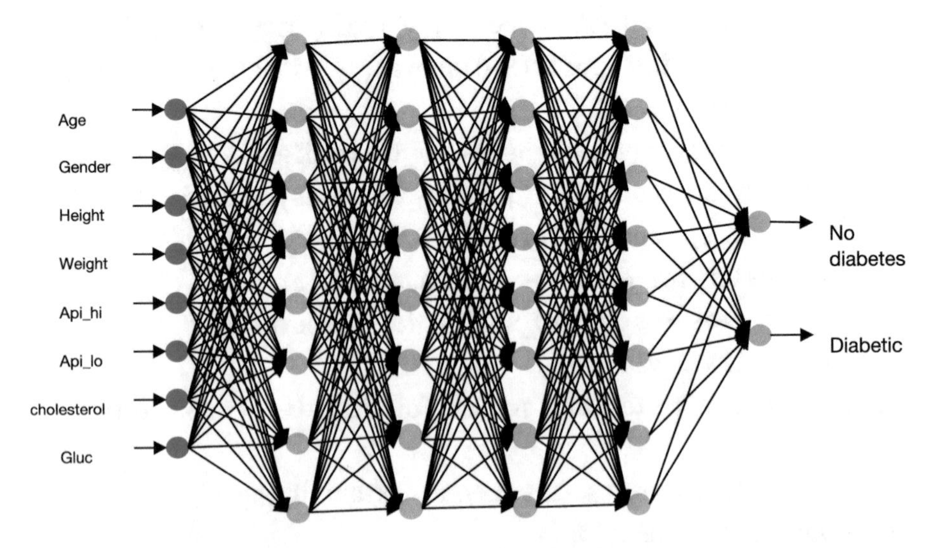

Figure 2-6. *The deep belief network's structure*

Outlining the Deep Belief Network's Predictions

Listing 2-16 outlines the deep belief network's predictions (see Table 2-4).

Listing 2-16. Outlining the Deep Belief Network's Predictions

```
y_hat_diabetes_dbn_model  = diabetes_dbn_model.predict(x_test_
diabetes)
actual_diabetes = pd.DataFrame(y_test_diabetes)
actual_diabetes.columns = ["Actual"]
predicted_diabetes = pd.DataFrame(y_test_diabetes)
predicted_diabetes.columns = ["Predicted"]
actual_and_predicted_diabetes = pd.concat([actual_diabetes,
predicted_diabetes], axis = 1)
actual_and_predicted_diabetes.loc[actual_and_predicted_
diabetes.Actual == 0, "Actual"] = "Not diabetic"
actual_and_predicted_diabetes.loc[actual_and_predicted_
diabetes.Actual == 1, "Actual"] = "Diabetic"
actual_and_predicted_diabetes.loc[actual_and_predicted_
diabetes.Predicted == 0, "Predicted"] = "Not diabetic"
actual_and_predicted_diabetes.loc[actual_and_predicted_
diabetes.Predicted == 1, "Predicted"] = "Diabetic"
actual_and_predicted_diabetes
```

Table 2-4. The Deep Belief Network's Predictions

	Actual	Predicted
0	Diabetic	Diabetic
1	Not diabetic	Not diabetic
2	Not diabetic	Not diabetic
3	Diabetic	Diabetic
4	Not diabetic	Not diabetic
...	...	...
149	Diabetic	Diabetic
150	Not diabetic	Not diabetic
151	Diabetic	Diabetic
152	Not diabetic	Not diabetic
153	Not diabetic	Not diabetic

Considering the Deep Neural Network's Performance

To determine how well the deep neural network classifies patient diabetes outcomes in training and cross-validation, this segment monitors the degree of binary cross-entropy loss and accuracy metric fluctuations as epochs increase. First, let's look at the confusion matrix and then the classification report. See Listing 2-17.

Listing 2-17. Outlining the Deep Belief Network's Confusion Matrix

```
diabetes_dbn_model_confusion_matrix = pd.DataFrame(confusion_
matrix(y_test_diabetes, y_hat_diabetes_dbn_model),
                      index=["Actual: No diabetes",
                         "Actual: Diabetic"],
                      columns = ("Predicted: No diabetes",
                           "Predicted: Diabetic"))
diabetes_dbn_model_confusion_matrix
```

Table 2-5. *The Deep Belief Network's Confusion Matrix*

	Predicted: No diabetes	Predicted: Diabetic
Actual: No diabetes	90	17
Actual: Diabetic	17	30

Table 2-5 shows that the deep belief network

- Correctly predicted patients had no diabetes 90 times.
- Incorrectly predicted that patients had diabetes when they did not have it 17 times.
- Incorrectly predicted that patients had no cardiovascular disease when they did have it 17 times.
- Correctly predicted patients had diabetes 30 times.

Listing 2-18 outlines the neural network's classification report, which holds the accuracy score, precision score, recall, f1 score, and support (see Table 2-6).

Listing 2-18. Outlining the Deep Belief Network's Classification Report

```
diabetes_dbn_model_report = pd.DataFrame(classification_
report(y_test_diabetes, y_hat_diabetes_dbn_model,
                              output_dict = True)).transpose()
diabetes_dbn_model_report
```

Table 2-6. *The Deep Belief Network's Classification Report*

	Precision	Recall	F-1 score	Support
0	0.841121	0.841121	0.841121	107.000000
1	0.638298	0.638298	0.638298	47.000000
Accuracy	0.779221	0.779221	0.779221	0.779221
Macro avg	0.739710	0.739710	0.739710	154.000000
Weighted avg	0.779221	0.779221	0.779221	154.000000

Table 2-6 suggests that the deep belief network is precise 84% of the time when predicting the absence of cardiovascular disease and 64% precise when predicting its presence. The overall accuracy score is 78%.

Accuracy Fluctuations Across Epochs in Training and Cross-Validation

Figure 2-7 depicts the degree of accuracy fluctuations as epochs increase in training and cross-validation when the deep neural network classifies patient diabetes outcomes. See Listing 2-19 for the code.

Listing 2-19. Charting Accuracy Fluctuations Across Epochs in Training and Cross-Validation

```
plt.plot(diabetes_dbn_model_history.history["accuracy"],
         color="orange",
         marker = "o",
         label = "Training accuracy")
plt.plot(diabetes_dbn_model_history.history["val_accuracy"],
         color="blue",
         marker = "o",
         label = "CV accuracy")
plt.xlabel("Epochs")
plt.ylabel("Loss")
plt.legend(loc="best")
plt.show()
```

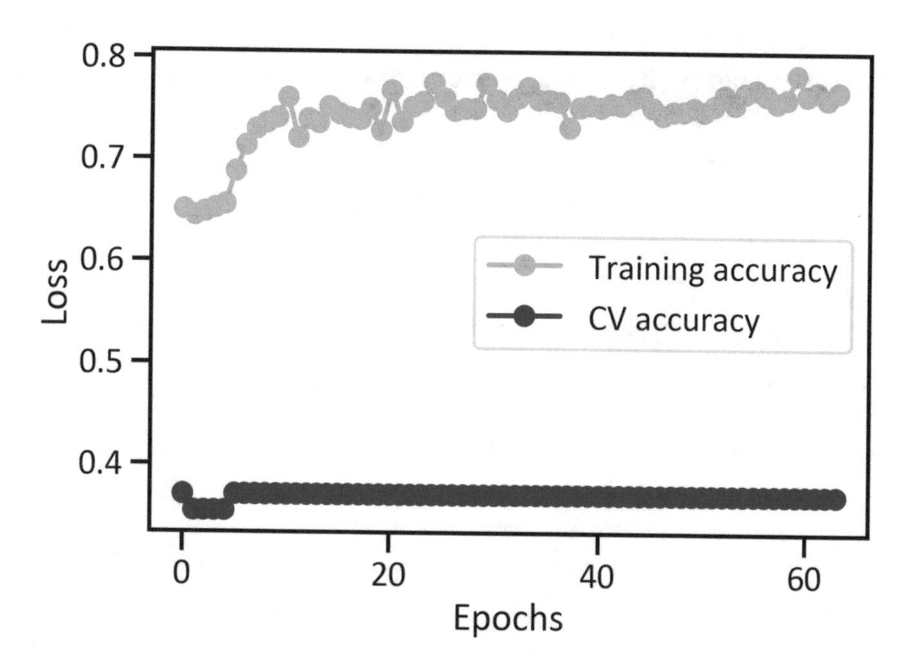

Figure 2-7. *Accuracy fluctuations across epochs in training and cross-validation*

Figure 2-7 shows that the accuracy is higher in training than cross-validation when classifying diabetes outcomes.

Binary Cross-Entropy Loss Fluctuations Across Epochs in Training and Cross-Validation

Listing 2-20 and Figure 2-8 depict the degree of binary cross-entropy loss fluctuations as epochs increase in training and cross-validation when the deep neural network classifies patient diabetes outcomes.

Listing 2-20. Binary Cross-Entropy Loss Fluctuations Across Epochs in Training and Cross-Validation

```
plt.plot(diabetes_dbn_model_history.history["loss"],
         color = "orange",
         marker = "o",
         label = "Training loss")
plt.plot(diabetes_dbn_model_history.history["val_loss"],
         color = "blue",
         marker = "o",
         label = "CV loss")
plt.xlabel("Epochs")
plt.ylabel("Loss")
plt.legend(loc = "best")
plt.show()
```

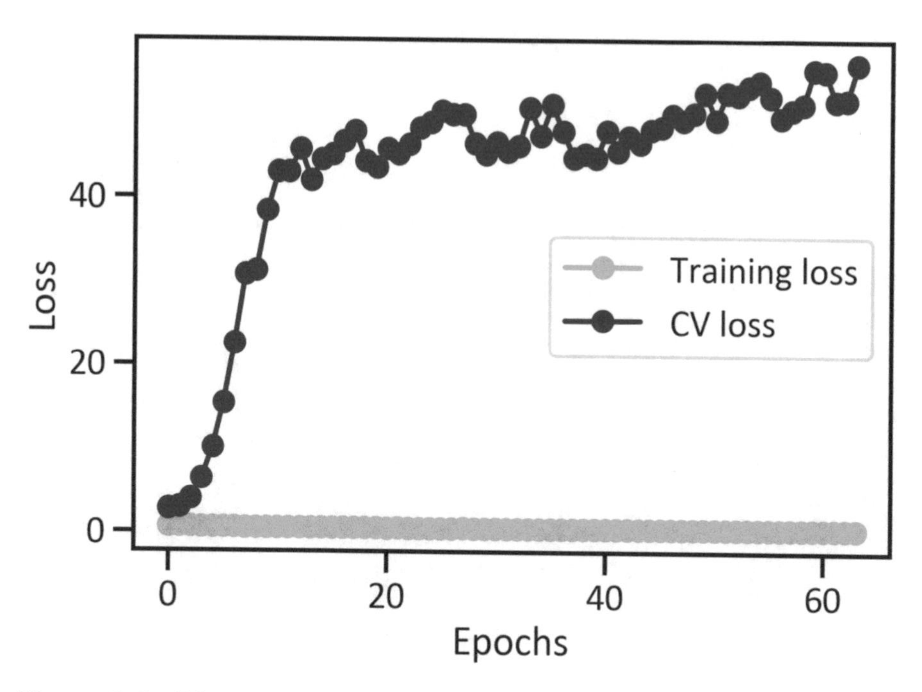

Figure 2-8. *Binary cross-entropy loss fluctuations across epochs in training and cross-validation*

Figure 2-8 shows that the binary cross-entropy loss is constant in training, but surges upwards in cross-validation.

Conclusion

This chapter marked the practical application of artificial neural networks for disease realization and segmentation. To begin, it summarized two preventable diseases. Subsequently, it provided some necessary background on neural networks. Then, it presented an approach for designing the structure of the network, training it, and ascertaining the performance.

CHAPTER 3

A Case for COVID-19: Considering the Hidden States and Simulation Results

In this chapter, you'll carry out a set of sequential methods (or time series methods) for forecasting to discern patterns in confirmed COVID-19 cases in the US. To begin, you'll use the Gaussian Hidden Markov Model to inherit the series, model it, and consider the hidden states, including the means and covariance in those states. Subsequently, you'll use the Monte Carlo simulation method to replicate confirmed US COVID-19 cases across multiple trials, therefore providing a rich comprehension of patterns in the data.

Executing the Hidden Markov Model

You start by using the Hidden Markov Model to uncover hidden states, acknowledged as classes or categorizes, in US-related confirmed COVID-19 cases. You'll employ the most simplified Markov Model identified as the Gaussian Hidden Markov Model, which returns two states 0 (signifying increasing US confirmed COVID-19 cases) and 1 (signifying decreasing US confirmed COVID-19 cases).

T. C. Nokeri, *Artificial Intelligence in Medical Sciences and Psychology*,
https://doi.org/10.1007/978-1-4842-8217-5_3

Let's base the Hidden Markov Model on the premise that current and subsequent states rely on preceding states. Equation 3-1 defines the Gaussian Hidden Markov model as

$$P\left(,S_{i2},S_{i3}\ldots,S_{ik-1}\right)=P\left(S_{ij}|S_{ik-1}\right) \quad \text{(Equation 3-1)}$$

where S_{ij} delineates the independent observation of the hidden states, S_{i1} delineates the first hidden state, S_{i2} delineates the second hidden state, and so forth.

Figure 3-1 shows the transition probabilities (defined in Equation 3-2) and initial probabilities (defined in Equation 3-3):

$$\alpha_{ij}=P\left(S_i|S_j\right) \quad \text{(Equation 3-2)}$$

and

$$\pi_i=P\left(S_i\right). \quad \text{(Equation 3-3)}$$

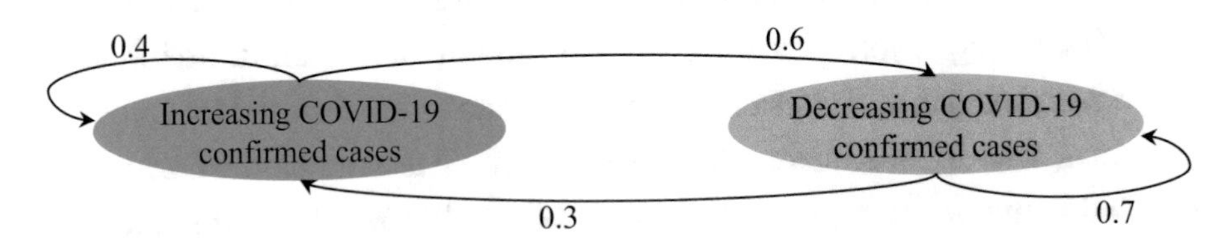

***Figure 3-1.** Transition probabilities*

Figure 3-1 shows two states, `"increasing confirmed COVID-19 cases"` and `"decreasing confirmed COVID-19 cases,"` whereby the transition probabilities are `P("increasing confirmed COVID-19 cases" | "increasing confirmed COVID-19 cases")` = 0.4 and `P("decreasing confirmed COVID-19 cases" | "increasing confirmed COVID-19 cases")` = 0.6.

The data that Johns Hopkins University collected was downloaded on September 24, 2021, from GitHub[1].

Note Johns Hopkins University updates the data daily. To reproduce the exercise, download the dataset from the GitHub folder for this chapter.

Descriptive Analysis

Before you execute the Gaussian Hidden Markov Model, first you should perform exploratory descriptive analysis. In this section, you consider the distribution of the data by constructing a histogram and a box plot. Following that, you describe the data by its central tendency and dispersion. Listing 3-1 collects the data and identifies null values (see Figure 3-2). Start by installing pandas in your environment: `pip install pandas`. In addition, install Matplotlib in your environment: `pip install matplotlib`. Also, install seaborn in your environment: `pip install seaborn`.

Listing 3-1. Collecting Data on US Confirmed COVID-19 Cases

```
import pandas as pd
import matplotlib.pyplot as plt
%matplotlib inline
import seaborn as sns
```

[1] https://github.com/CSSEGISandData/COVID-19/blob/master/csse_covid_19_data/csse_covid_19_time_series/time_series_covid19_deaths_US.csv

```
covid_us_df = pd.read_csv(r"filepaht\time_series_covid19_
deaths_US.csv", index_col=[0], parse_dates=[0])["cases_
confirmed"]
covid_us_df = pd.DataFrame(covid_us_df)
sns.heatmap(covid_us_df.isnull())
plt.show()
```

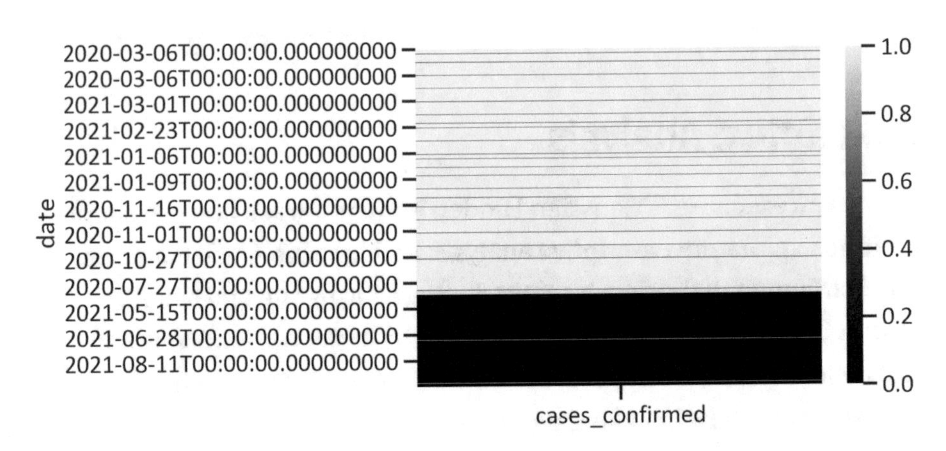

Figure 3-2. *Heatmap for null values*

Listing 3-2 drops missing values. In addition, it plots another heatmap to verify whether null values were dropped (see Figure 3-3).

Listing 3-2. Substituting Null Values

```
import seaborn as sns
covid_us_df = covid_us_df.dropna()
sns.heatmap(covid_us_df.isnull())
plt.show()
```

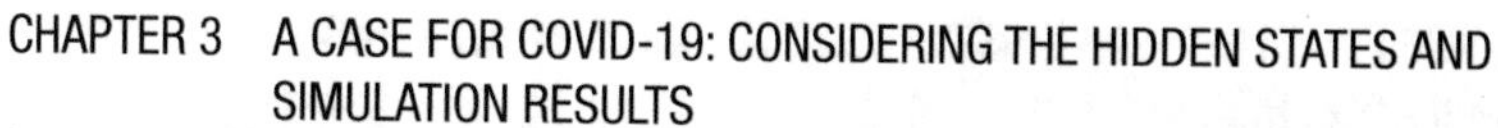

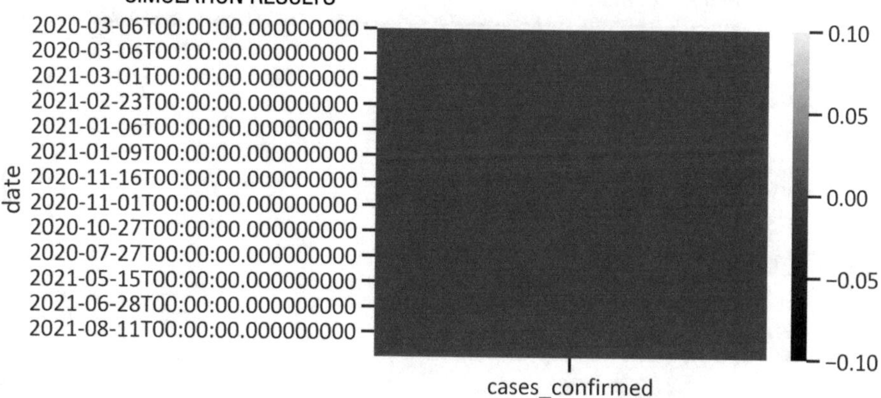

Figure 3-3. *Heatmap for null values*

The most suitable approach to studying the inclination of the data towards central instances involves constructing the confidence interval against the frequency of the data using a histogram.

Listing 3-3 constructs a histogram of US confirmed COVID-19 cases (see Figure 3-4).

Listing 3-3. US Confirmed COVID-19 Cases Histogram

```
sns.histplot(data=covid_us_df, x = covid_us_df.cases_confirmed,
color = "orange")
plt.show()
```

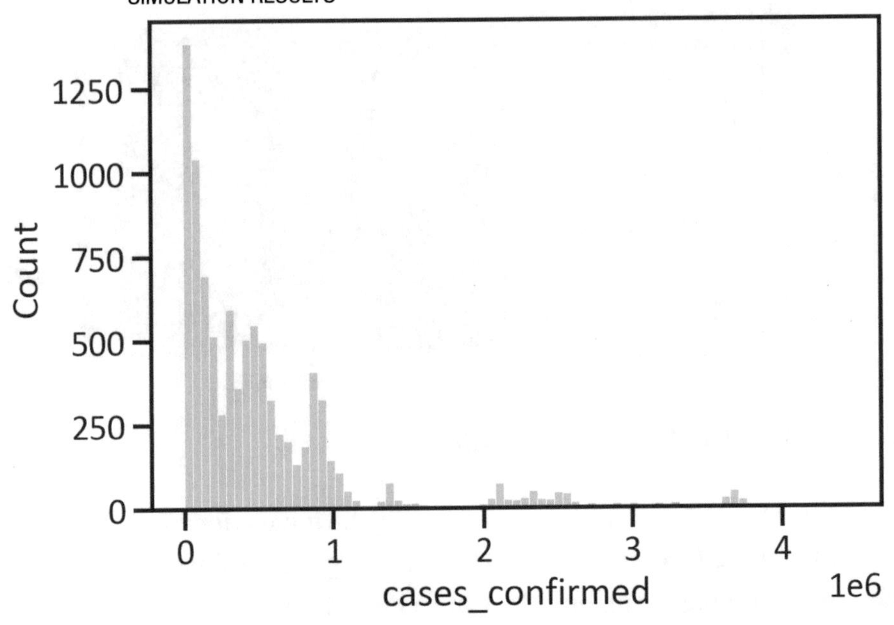

Figure 3-4. *US confirmed COVID-19 cases*

Listing 3-4 constructs a box plot of US confirmed COVID-19 cases to confirm the distribution found in Figure 3-3 (see Figure 3-5).

Listing 3-4. US Confirmed COVID-19 Cases Box Plot

```
sns.boxplot(covid_us_df.cases_confirmed, color = "orange")
plt.show()
```

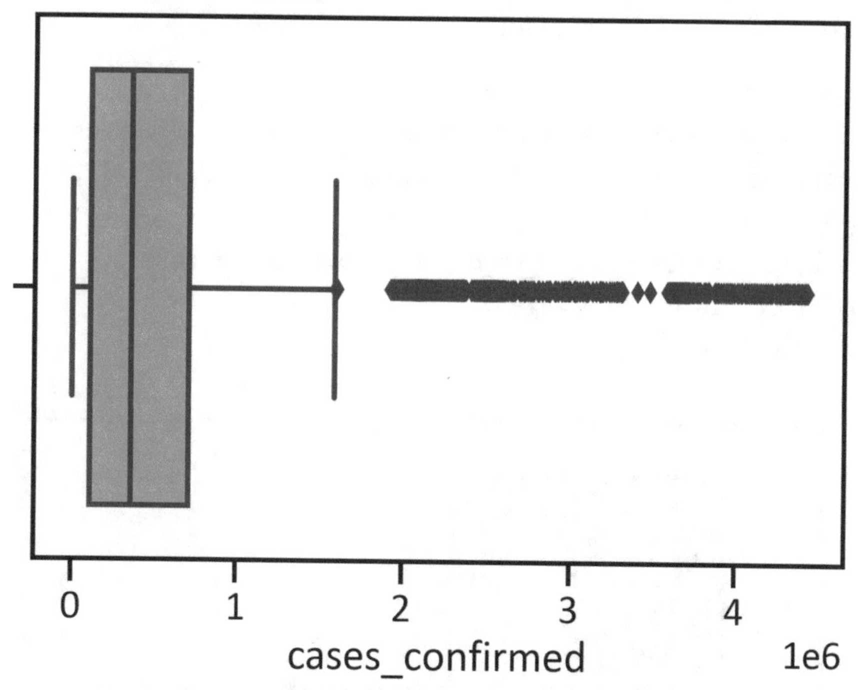

Figure 3-5. *US confirmed COVID-19 cases box plot*

Figure 3-5 confirms the inclination of US confirmed COVID-19 cases data towards the left. In addition, it indicates vast outliers.

To prevent outliers from misrepresenting the data, you perform imputation, which is the substitution of values using some specified value (in this case, you substitute outliers with the mean value).

Listing 3-5 substitutes outliers with the mean values. Then, it constructs another box plot to verify whether the outliers have been removed (see Figure 3-6). Start by installing NumPy in your environment: `pip install numpy`.

Listing 3-5. Substituting Outliers

```
import numpy as np
covid_us_df.cases_confirmed = np.where((covid_us_df.cases_
confirmed > 1.25e+6),covid_us_df.cases_confirmed .mean(),covid_
us_df.cases_confirmed)
sns.boxplot(data = covid_us_df, x = covid_us_df.cases_
confirmed, color = "orange")
plt.show()
```

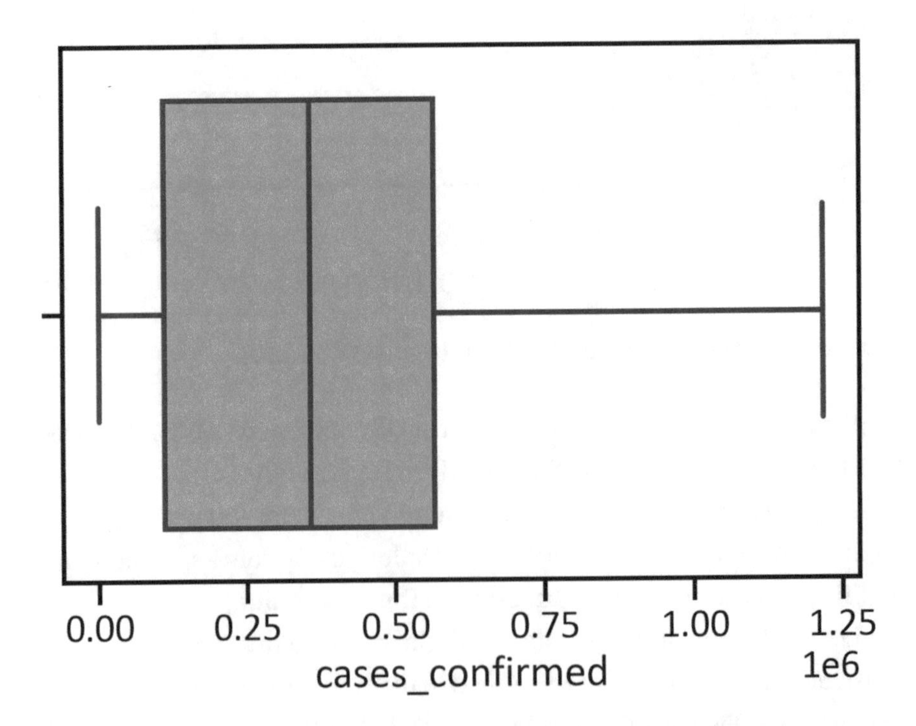

Figure 3-6. *US confirmed COVID-19 cases box plot after outlier removal*

Figure 3-6 confirms that the mean value indeed substituted outliers. Listing 3-6 plots US confirmed COVID-19 cases series (Figure 3-7).

Listing 3-6. Plotting US Confirmed COVID-19 Cases Series

```
sns.lineplot(data=covid_us_df, x = covid_us_df.index, y =
covid_us_df.cases_confirmed, color = "orange")
plt.xticks (rotation = 90)
plt.show()
```

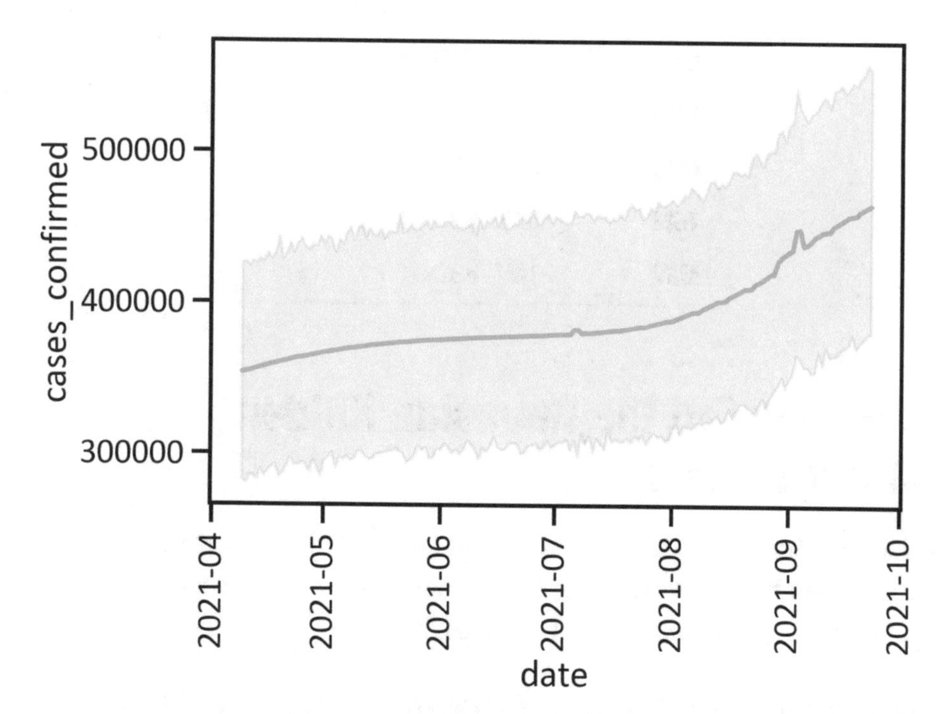

Figure 3-7. *US confirmed COVID-19 cases series*

Listing 3-7 reports on the mean, median, and standard deviation, among other key statistics (see Table 3-1).

Listing 3-7. Computing Descriptive Statistics

```
covid_us_df.describe()
```

Table 3-1. Descriptive Statistics

	cases_confirmed
Count	9.380000e+03
Mean	3.900083e+05
Std	2.952401e+05
Min	0.000000e+00
25%	1.094280e+05
50%	3.560650e+05
75%	5.653697e+05
Max	1.217892e+06

Carrying Out the Gaussian Hidden Markov Model

Listing 3-8 carries out the Gaussian Hidden Markov Model with `n_components` as 2 across 10 iterations. Start by installing hmmlearn in your environment: `pip install hmmlearn`.

Listing 3-8. Carrying Out the Gaussian Hidden Markov Model

```
from hmmlearn.hmm import GaussianHMM
hmm_data = np.column_stack([covid_us_df])
gaussian_hmm_model = GaussianHMM(n_components=2, tol=0.0001,
n_iter=10)
gaussian_hmm_model.fit(hmm_data)
```

Considering the Hidden States in US Confirmed COVID-19 Cases with the Gaussian Hidden Markov Model

Listing 3-9 considers the hidden states in US confirmed COVID-19 cases with the Gaussian Hidden Markov Model (Table 3-2).

Listing 3-9. Considering the Hidden States in US Confirmed COVID-19 Cases with the Gaussian Hidden Markov Model

```
hidden_states = pd.DataFrame(gaussian_hmm_model.predict
(hmm_data), columns = ["hidden_states"])
hidden_states.index = covid_us_df.index
hidden_states.head()
```

Table 3-2. *Hidden States in US Confirmed COVID-19 Cases with the Gaussian Hidden Markov Model*

	hidden_states
Date	
2021-04-09	1
2021-04-10	1
2021-04-11	1
2021-04-12	1
2021-04-13	1

Listing 3-10 describes the Gaussian Hidden Markov Model hidden states (see Table 3-3).

Listing 3-10. Gaussian Hidden Markov Model Hidden States Descriptive Statistics

```
hidden_states.describe()
```

Table 3-3. *Gaussian Hidden Markov Model Hidden States Descriptive Statistics*

	hidden_states
Count	9380.000000
Mean	0.430384
Std	0.495156
Min	0.000000
25%	0.000000
50%	0.000000
75%	1.000000
Max	1.000000

Table 3-3 shows that the mean value is 0.43 and the standard deviation is 0.49.

Listing 3-11 depicts the Gaussian Hidden Markov Model hidden states (Figure 3-8).

Listing 3-11. Depicting the Gaussian Hidden Markov Model Hidden States

```
n_sample = 500
sample, _ = gaussian_hmm_model.sample(n_sample)
plt.plot(np.arange(n_sample), sample[:,0], color = "orange")
plt.xlabel("Samples")
plt.ylabel("States")
plt.show()
```

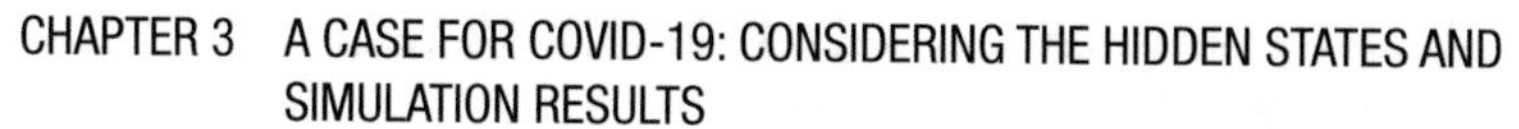

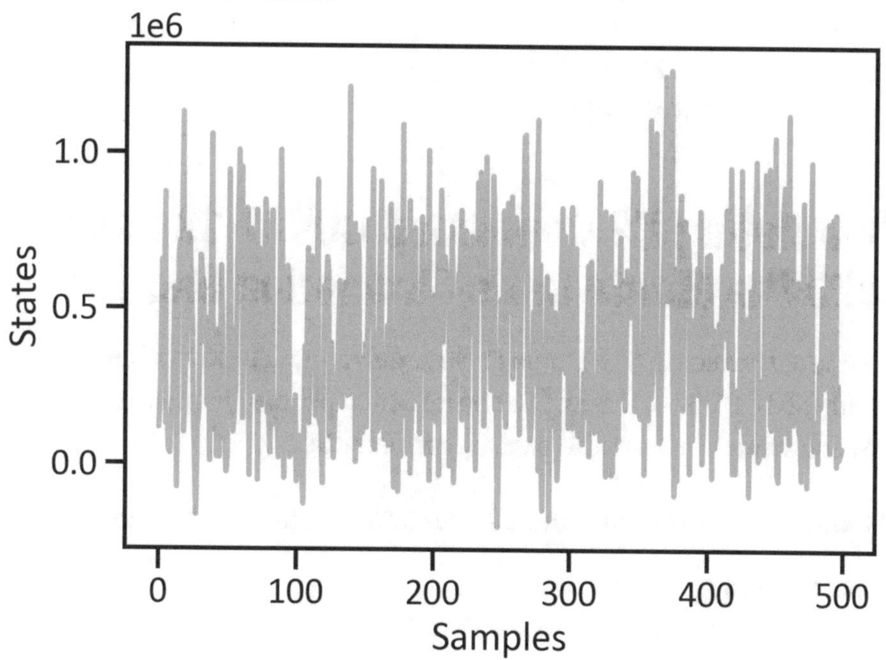

Figure 3-8. *Gaussian Hidden Markov Model hidden states*

Figure 3-8 shows that the 500 samples are mostly between 0 and 1, with minor exceptions.

Listing 3-12 identifies the means and covariances of the Gaussian Hidden Markov Model's hidden states.

Listing 3-12. Identifying the Hidden States' Means and Covariances

```
for i in range(gaussian_hmm_model.n_components):
    print("{0} order hidden state".format(i))
    print("mean = ", gaussian_hmm_model.means_[i])
    print("var = ", np.diag(gaussian_hmm_model.covars_[i]))
    print()
0 order hidden state
mean =  [591862.57352145]
var =  [5.03979663e+10]
```

```
1 order hidden state
mean =  [121584.27846397]
var =  [9.80585646e+09]
```

Simulating US Confirmed COVID-19 Cases with the Monte Carlo Simulation Method

Listing 3-13 executes the Monte Carlo simulation method to simulate US confirmed COVID-19 cases. Start by installing pandas_montecarlo in your environment: `pip install pandas_montecarlo`.

Listing 3-13. Executing the Monte Carlo Simulation Method

```
import pandas_montecarlo
monte_carlo_model = covid_us_df.cases_confirmed.
montecarlo(sims=5, bust=-0.1, goal=1)
```

US Confirmed COVID-19 Cases Simulation Results

Listing 3-14 depicts the Monte Carlo simulation results of US confirmed COVID-19 cases (see Figure 3-9).

Listing 3-14. Depicting the Monte Carlo Simulation Results of US Confirmed COVID-19 Cases

```
monte_carlo_model.plot(title="")
```

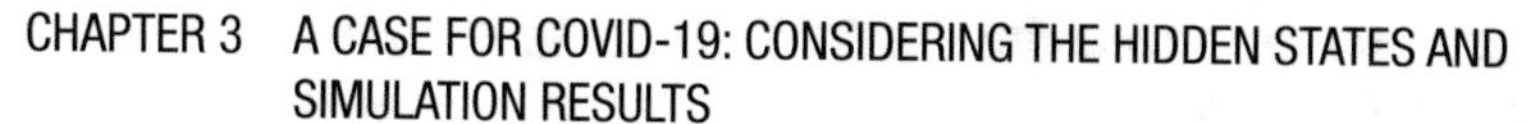

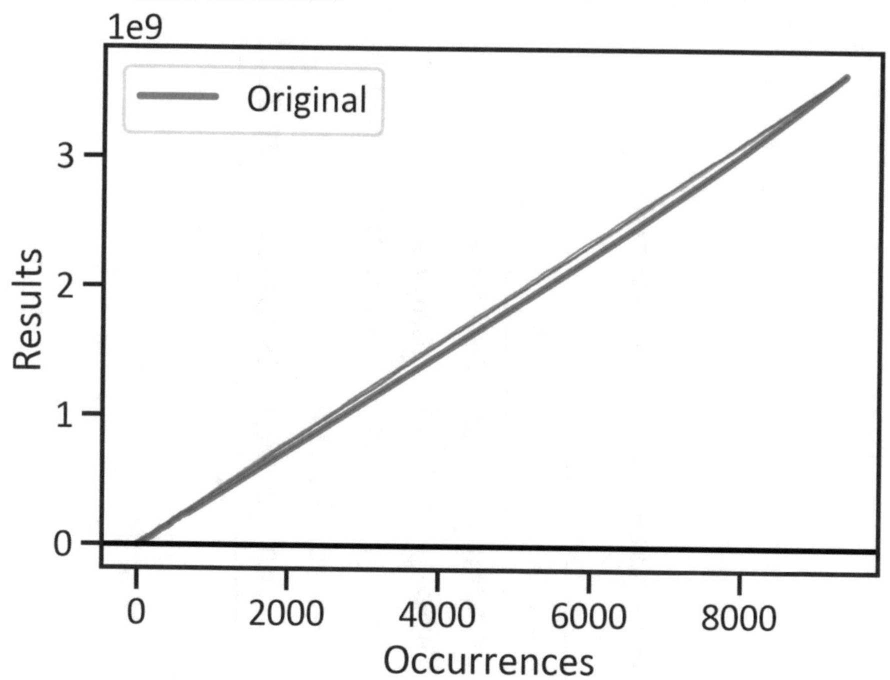

Figure 3-9. *Monte Carlo simulation results of US confirmed COVID-19 cases*

Listing 3-15 outlines the descriptive statistics for the Monte Carlo simulation results of US confirmed COVID-19 cases (see Table 3-4).

Listing 3-15. Descriptive Statistics for the Monte Carlo Simulation Results of US Confirmed COVID-19 Cases

```
monte_carlo_model_output = pd.DataFrame(monte_carlo_model.data)
monte_carlo_model_output.describe().transpose()
```

Table 3-4. *Descriptive Statistics for Monte Carlo's Simulation Results of US Confirmed COVID-19 Cases*

	Count	Mean	Std	Min	25%	50%	75%	Max
Original	9380.0	390008.25521	295240.11557	0.0	109428.0	356065.0	565369.72484	1217892.0
1	9380.0	390008.25521	295240.11557	0.0	109428.0	356065.0	565369.72484	1217892.0
2	9380.0	390008.25521	295240.11557	0.0	109428.0	356065.0	565369.72484	1217892.0
3	9380.0	390008.25521	295240.11557	0.0	109428.0	356065.0	565369.72484	1217892.0
4	9380.0	390008.25521	295240.11557	0.0	109428.0	356065.0	565369.72484	1217892.0

Listing 3-16 constructs the descriptive statistics for the Monte Carlo simulation results of US confirmed COVID-19 cases (see Figure 3-10).

Listing 3-16. Constructing a Histogram for the Monte Carlo Simulation Results of US Confirmed COVID-19 Cases

```
fig, ax = plt.subplots(figsize= (12, 7))
sns.boxplot(data = monte_carlo_model_simulation_results,
color = "orange")
plt.ylabel("Values")
plt.show()
```

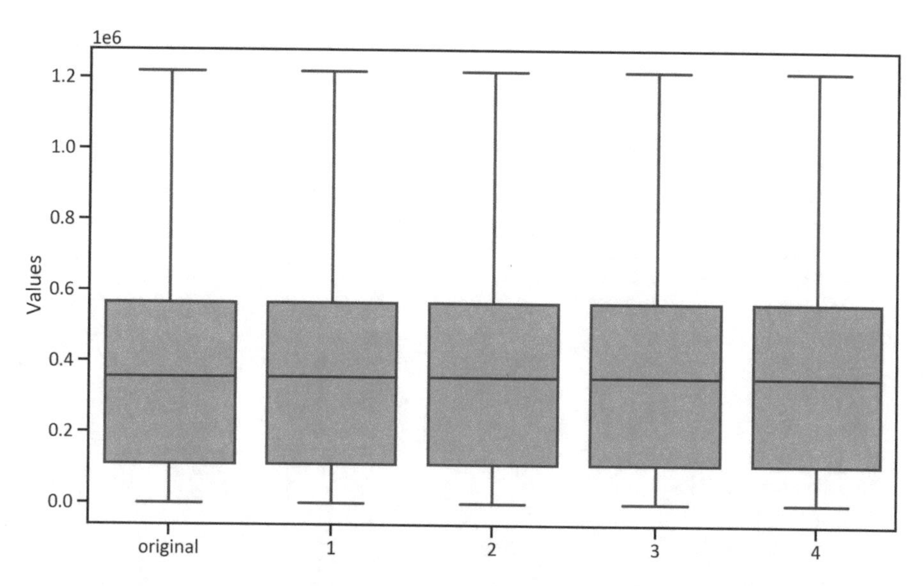

Figure 3-10. *Histogram for the Monte Carlo simulation results of US confirmed COVID-19 cases*

Conclusion

This chapter shows two methods for remedying sequential problems, the Gaussian Hidden Markov Model and the Monte Carlo simulation. The Markov model returned two states, 0 (signifying increasing US confirmed COVID-19 cases) and 1 (signifying decreasing US confirmed COVID-19 cases). Following that, it carried out five simulation trials with the Monte Carlo simulation method, all unveiling rising cases.

CHAPTER 4

Cancer Segmentation with Neural Networks

This chapter explains the practical application of computer vision and convolutional neural networks for breast and skin cancer realization and segmentation. In it, you'll explore an approach to filtering medical scans by applying `canny`, `luplican`, and `sobel` filters. You'll also ascertain the extent to which the networks accurately differentiate scans of patients with and without cancer.

Exploring Cancer

Cancer is a disease that marks an irregular growth of cells with an inclination to spread transversely across the body. We recognize this specific disease by the visible presence of a tumor, atypical bleeding, and a prolonged cough.

There is a vast pool of cancer-causing agents. Besides exposure to certain chemicals, lifestyle behaviors such as smoking and excessive alcohol consumption may be risks.

The most prevalent approaches for diagnosing cancer in a patient involve MRI and ultrasound scans. In its developmental stages, cancer can be treated by chemotherapy, surgery, and radiation therapy, among others. Provided the above, it is crucial to conduct periodic medical checks to diagnose cancer.

T. C. Nokeri, *Artificial Intelligence in Medical Sciences and Psychology*,
https://doi.org/10.1007/978-1-4842-8217-5_4

Exploring Skin Cancer

Let's execute a convolutional neural network (CNN) to realize the visible presence of skin cancer. Figure 4-1 depicts the dominant forms of skin cancer that the network will differentiate.

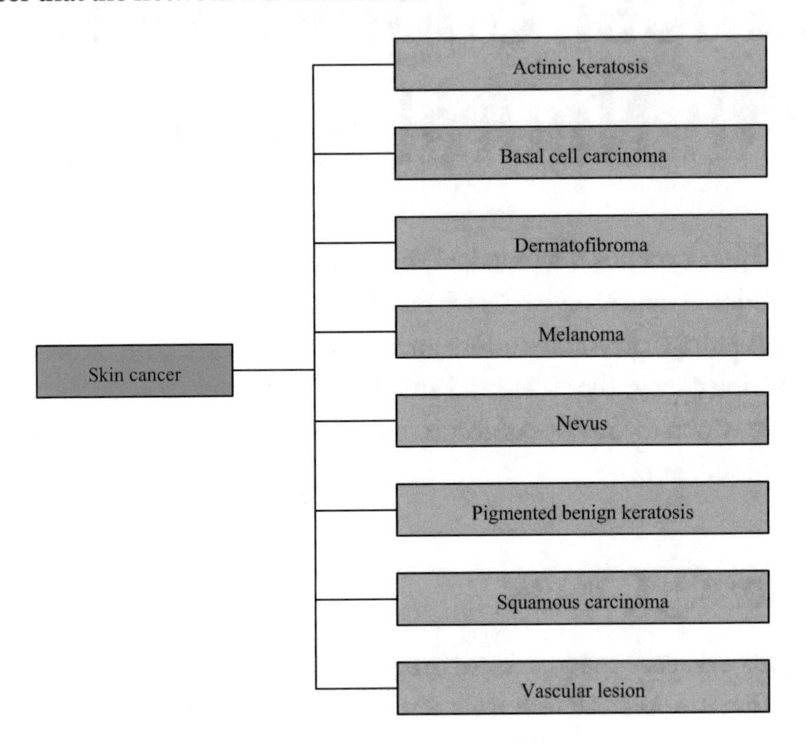

Figure 4-1. *Varyious forms of skin cancer*

Figure 4-1 shows common forms of skin cancer such as actinic keratosis, basal cell carcinoma, dermatofibroma, and more.

Classifying Patient Skin Cancer Outcomes by Executing a CNN

Convolutional neural networks are a special case of networks widely applied in image classification. They are a perfect candidate for the use cases in this chapter. You will apply a machine learning algorithm that learns patterns in existing skin cancer data, feed it with unseen skin cancer scan data, and conclude how well it classifies the scans. Figure 4-2 depicts a simple example of a convolutional neural network.

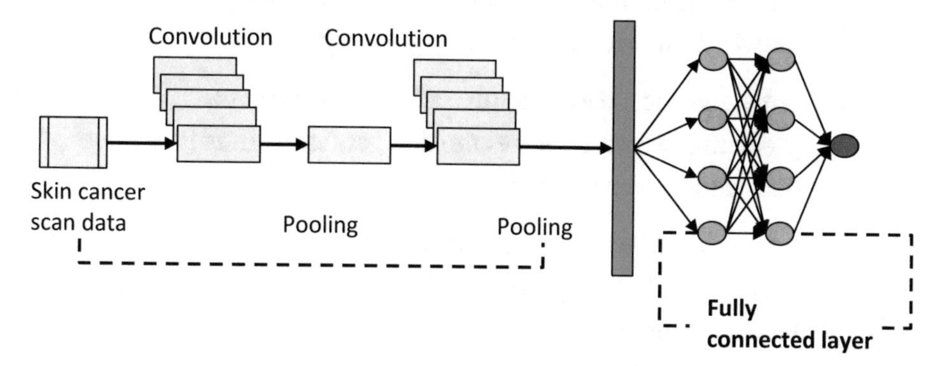

Figure 4-2. *Example of a convolutional neural network*

Figure 4-2 depicts a CNN comprising an input layer that receives input skin cancer scans. Following that, it reduces the input data into a few dimensions to form parts of an object using a linear or non-linear kernel. Subsequently, it aggregates the reduced data to form the object. This object detection procedure replicates that of the human visual cortex.

At most, a CNN outperforms a shallow neural network because of its ability to conduct dimension reduction. As seen in Figure 4-2, a CNN comprises these key layers:

- **Convolutional layer:** Receives data from the input layer. Then it identifies and prioritizes important features in the data and transmits the data into the pooling layer.
- **Pooling layer:** Processes and aggregates the features received from the convolutional layer into a single individual neuron in the subsequent layer.
- **Fully connected layer:** Links all neurons from the preceding layers, operates the data, and transmits the data into the output layer.

Besides the use cases in this chapter, a CNN can be used in face detection, recommendation systems, and natural language processing, among other applications.

A CNN Pipeline

This chapter follows the simple pipeline depicted in Figure 4-3. A pipeline is steps that sum up the model lifecycle.

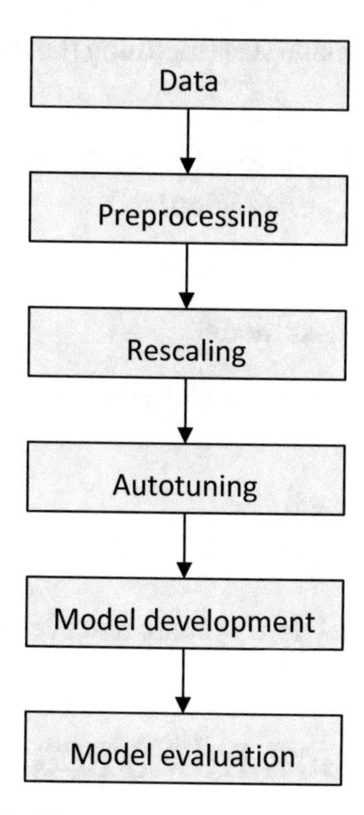

Figure 4-3. *A CNN pipeline*

Figure 4-3 depicts data extraction as the initial step in the pipeline. The data used in this chapter was extracted from Kaggle[1]. The subsequent step is data preprocessing, which involves the following tasks:

- Specifying data subsets for training and validation (setting aside a portion of data that the CNN learns image patterns from and another portion used to validate its estimates)
- Specifying a validation split (the ratio for splitting the data into training and validation data)

[1] `https://www.kaggle.com/datasets/nodoubttome/skin-cancer9-classesisic`

- Assigning the image size (inputting the image channels and pixels)
- Specifying the batch size (the number of training labels to be processed per iteration)

The next step is data rescaling, which involves these tasks:

- Specifying the class mode
- Specifying the target size (the size of images that the CNN generates)
- Assigning the batch size
- Generating labels

Following this is model development and evaluation.

A CNN's Architectural Structure

The CNN that classifies patient brain tumor outcomes has a relatively simple architectural structure, as shown in Figure 4-4.

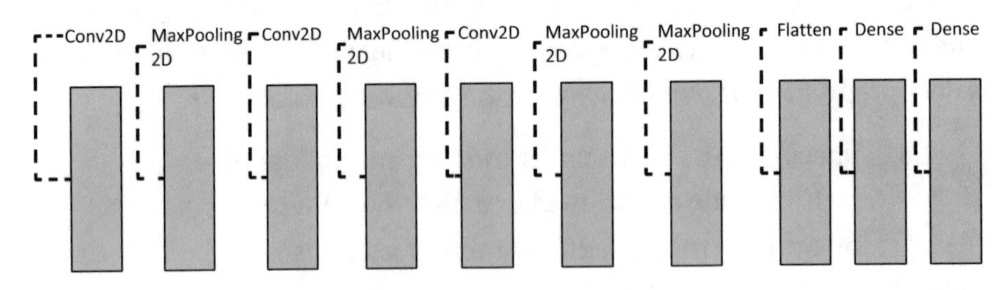

Figure 4-4. *A CNN's architectural structure*

The CNN comprises a set of 2D convolutional and MaxPooling2D layers, a flatten layer, and two dense layers, with all layers consisting of a rectified linear unit (`relu`) activation function that absorbs and models inputs to estimate a set of unconstrained values of the target feature. Chapter 2 discusses the `relu` function in detail.

Classifying Skin Cancer Diagnosis Image Data by Executing a CNN

Listing 4-1 collects the data by executing the `pathlib` library.

Listing 4-1. Collecting the Skin Cancer Diagnosis Image Data

```
import pathlib
skin_cancer_data = pathlib.Path(r"filepath")
```

Preprocessing the Training Skin Cancer Image Data

Listing 4-2 preprocesses the images (for training) by executing the `image_dataset_from_directory()` method in the `keras` library. Equally, it outlines the `validation_split`, `subset`, `image_size`, and `batch_size`.

Listing 4-2. Preprocessing the Training Skin Cancer Image Data

```
skin_cancer_training_data = tf.keras.preprocessing.
image_dataset_from_directory(skin_cancer_data, seed = 123,
validation_split = 0.2, subset = "training",  image_size =
(180, 180), batch_size = 16)
```

Preprocessing the Validation Skin Cancer Image Data

Listing 4-3 preprocesses the images (for validation) by executing the `image_dataset_from_directory()` method in the keras library. Equally, it outlines the `validation_split`, `subset`, `image_size`, and `batch_size`.

Listing 4-3. Preprocessing the Validation Skin Cancer Image Data

```
skin_cancer_validation_data = tf.keras.preprocessing.
image_dataset_from_directory(skin_cancer_data, seed = 123,
validation_split = 0.2, subset = "validation", image_size =
(180, 180), batch_size = 16)
```

Generating the Training Skin Cancer Diagnosis Image Data

Listing 4-4 generates images or training by executing the `ImageDataGenerator()` method and outlining the rescaled ratio. Then, it executes the `flow_from_directory()` method and outlines the `target_size`, `class_mode`, and `batch_size`.

Listing 4-4. Generating the Training Skin Cancer Diagnosis Image Data

```
import keras
skin_cancer_training_data_categories = np.array(skin_cancer_
training_data.class_names)
skin_cancer_training_data_categories = np.array(skin_cancer_
training_data.class_names)
skin_cancer_generated_image_data = keras.preprocessing.image.
ImageDataGenerator(rescale = 1./255)
```

```
skin_cancer_generated_image_data_for_training = skin_cancer_
generated_image_data.flow_from_directory(skin_cancer_data,
                          target_size = (180, 180),
                          class_mode = "categorical",
                          shuffle = True,
                          batch_size = 16)
skin_cancer_images, skin_cancer_labels = next(iter(skin_cancer_
generated_image_data_for_training))
```

Tuning the Training Skin Cancer Image Data

This section develops a convolutional neural network. Initially, it prepares the image data in such a way that the network better discovers patterns by tuning (see Listing 4-5).

Listing 4-5. Tuning the Training Skin Cancer Image Data

```
skin_cancer_experimental_tuning = tf.data.experimental.AUTOTUNE
skin_cancer_training_data = skin_cancer_training_data.
cache().shuffle(1000).prefetch(buffer_size = skin_cancer_
experimental_tuning)
skin_cancer_validation_data = skin_cancer_validation_data.
cache().prefetch(buffer_size = skin_cancer_experimental_tuning)
```

Executing the CNN to Classify Skin Cancer Diagnosis Image Data

Listing 4-6 trains the convolutional neural network on the image data.

Listing 4-6. Executing the CNN to Classify Skin Cancer Diagnosis Image Data

```
from tensorflow.keras import layers
from tensorflow.python.keras.layers import Dense, Flatten,
Conv2D, Dropout, MaxPooling2D
skin_cancer_convolutional_net_model = Sequential([
    layers.experimental.preprocessing.Rescaling(1./255, input_
    shape = (180, 180, 3)),
    layers.Conv2D(16, 3, padding = "same", activation = "relu"),
    layers.MaxPooling2D(),
    layers.Conv2D(32, 3, padding = "same", activation = "relu"),
    layers.MaxPooling2D(),
    layers.Conv2D(64, 3, padding = "same", activation = "relu"),
    layers.MaxPooling2D(),
    layers.Flatten(),
    layers.Dense(128, activation = "relu"),
    layers.Dense(9, activation='softmax')])
skin_cancer_convolutional_net_model.compile(optimizer = "adam",
                                loss = tf.keras.losses.Sparse
                                CategoricalCrossentropy(from_
                                logits = True),
                                metrics = ["accuracy"])
skin_cancer_convolutional_net_model_history = skin_cancer_
convolutional_net_model.fit(skin_cancer_training_data,
                          validation_data = skin_cancer_
                          validation_data, epochs = 64)
skin_cancer_convolutional_net_model_history
```

Considering the CNN's Performance

To determine how well the CNN classifies patient skin cancer outcomes in training and cross-validation, this segment monitors the degree of sparse categorical cross-entropy loss and accuracy metric fluctuations as epochs increase.

Accuracy Fluctuations Across Epochs in Training and Cross-Validation

Figure 4-5 depicts the degree of accuracy fluctuations as epochs increase in training and cross-validation when the CNN classifies patient skin cancer outcomes. See Listing 4-7 for the code.

Listing 4-7. Charting Accuracy Fluctuations Across Epochs in Training and Cross-Validation

```
plt.plot(skin_cancer_convolutional_net_model_history.
history["accuracy"],
         color = "orange",
         marker = "o",
         label = "Training accuracy")
plt.plot(skin_cancer_convolutional_net_model_history.
history["val_accuracy"],
         color = "blue",
         marker = "o",
         label = "CV accuracy")
plt.xlabel("Epochs")
plt.ylabel("Accuracy")
plt.legend(loc = "best")
plt.show()
```

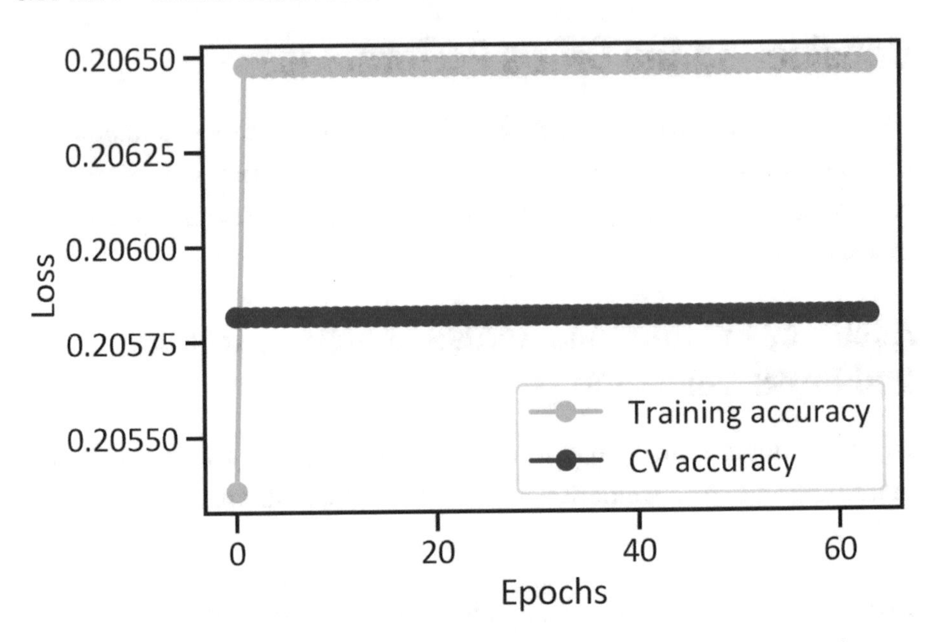

Figure 4-5. *Accuracy fluctuations across epochs in training and cross-validation*

Sparse Categorical Cross-Entropy Loss Fluctuations Across Epochs in Training and Cross-Validation

Figure 4-6 depicts the degree of sparse categorical cross-entropy loss fluctuations as epochs increase in training and cross-validation when the CNN classifies patient skin cancer outcomes. See Listing 4-8 for the code.

Listing 4-8. Charting Sparse Categorical Cross-Entropy Loss Fluctuations Across Epochs in Training and Cross-Validation

```
plt.plot(skin_cancer_convolutional_net_model_history.
history["loss"],
         color = "orange",
         marker = "o",
         label = "Training loss")
```

```
plt.plot(skin_cancer_convolutional_net_model_history.
history["val_loss"],
         color = "blue",
         marker = "o",
         label = "CV loss")
plt.xlabel("Epochs")
plt.ylabel("Loss")
plt.legend(loc = "best")
plt.show()
```

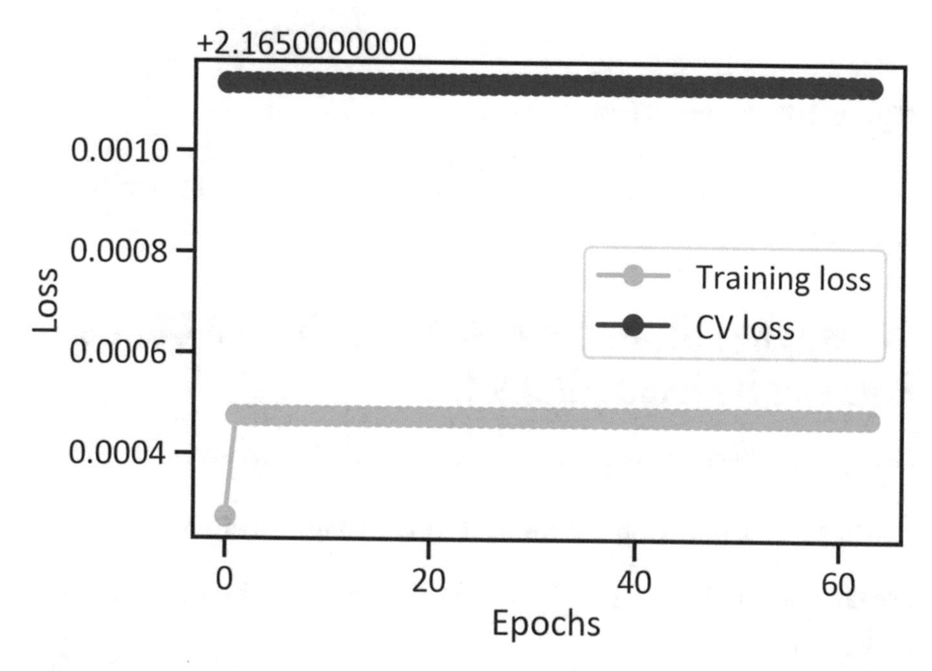

Figure 4-6. *Sparse categorical cross-entropy loss fluctuations across epochs in training and cross-validation*

Visible Presence of Breast Cancer

This section executes a CNN to realize the visible presence of breast cancer. Figure 4-7 depicts the main forms of breast cancer that the network will differentiate.

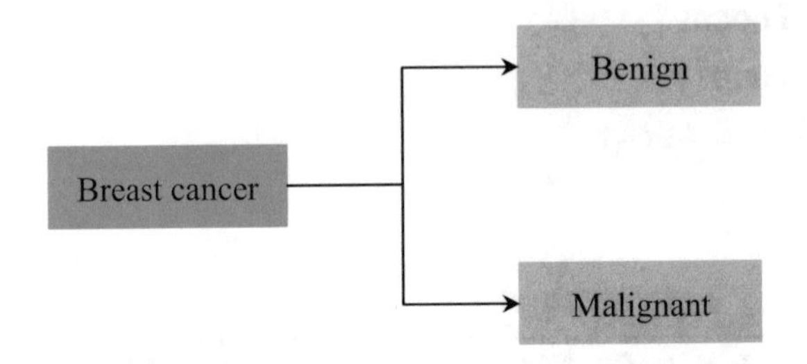

Figure 4-7. *Forms of breast cancer*

Figure 4-7 depicts two common forms of breast cancer: benign and malignant.

Classifying Ultrasound Scans of Breast Cancer Patients by Executing a CNN

Listing 4-9 collects the data by executing the `pathlib` library.

Listing 4-9. Collecting the Breast Cancer Diagnosis Image Data

```
breast_cancer_data = pathlib.Path(r"filepath\Dataset_BUSI_
with_GT")
```

Preprocessing the Validation Breast Cancer Image Data

Listing 4-10 preprocesses the images (for validation) by executing the `image_dataset_from_directory()` method in the keras library. Equally, it outlines the `validation_split`, `subset`, `image_size`, and `batch_size`.

Listing 4-10. Preprocessing the Validation Breast Cancer Image Data

```
breast_cancer_validation_data = tf.keras.preprocessing.
image_dataset_from_directory(breast_cancer_data, seed = 123,
validation_split = 0.2, subset = "validation", image_size =
(180, 180), batch_size = 16)
```

Generating the Training Breast Cancer Diagnosis Image Data

Listing 4-11 generates images or training by executing the `ImageDataGenerator()` method and outlining the rescaled ratio. Following that, it executes the `flow_from_directory()` method and outlines the `target_size`, `class_mode`, and `batch_size`.

Listing 4-11. Generating the Training Skin Cancer Diagnosis Image Data

```
breast_cancer_training_data_categories = np.array(breast_
cancer_training_data.class_names)
breast_cancer_training_data_categories = np.array(breast_
cancer_training_data.class_names)
breast_cancer_generated_image_data = keras.preprocessing.image.
ImageDataGenerator(rescale = 1./255)
```

```
breast_cancer_generated_image_data_for_training = breast_
cancer_generated_image_data.flow_from_directory(breast_
cancer_data, target_size = (180, 180), class_mode =
"categorical", shuffle = True, batch_size = 16)
breast_cancer_images, breast_cancer_labels = next(iter(breast_
cancer_generated_image_data_for_training))
```

Tuning the Training Breast Cancer Image Data

This section develops a CNN. Initially, it prepares the image data in such a way that the network better discovers patterns by tuning (see Listing 4-12).

Listing 4-12. Tuning the Training Breast Cancer Image Data

```
breast_cancer_experimental_tuning = tf.data.experimental.
AUTOTUNE
breast_cancer_training_data = breast_cancer_training_data.
cache().shuffle(1000).prefetch(buffer_size = breast_cancer_
experimental_tuning)
breast_cancer_validation_data = breast_cancer_validation_
data.cache().prefetch(buffer_size = breast_cancer_
experimental_tuning)
```

Executing the CNN to Classify Breast Cancer Diagnosis Image Data

Listing 4-13 trains the CNN on the image data.

Listing 4-13. Executing the CNN to Classify Breast Cancer Diagnosis Image Data

```
breast_cancer_convolutional_net_model = Sequential([
    layers.experimental.preprocessing.Rescaling(1./255,
    input_shape = (180, 180, 3)),
    layers.Conv2D(16, 3, padding = "same", activation = "relu"),
    layers.MaxPooling2D(),
    layers.Conv2D(32, 3, padding = "same", activation = "relu"),
    layers.MaxPooling2D(),
    layers.Conv2D(64, 3, padding = "same", activation = "relu"),
    layers.MaxPooling2D(),
    layers.Flatten(),
    layers.Dense(128, activation = "relu"),
    layers.Dense(9, activation='softmax')])
breast_cancer_convolutional_net_model.compile(optimizer =
"adam", loss = tf.keras.losses.SparseCategoricalCrossentropy
(from_logits = True), metrics = ["accuracy"])
breast_cancer_convolutional_net_model_history = breast_cancer_
convolutional_net_model.fit(breast_cancer_training_data,
validation_data = breast_cancer_validation_data, epochs = 64)
breast_cancer_convolutional_net_model_history
```

Considering the CNN's Performance

To determine how well the CNN classifies patient breast cancer outcomes in training and cross-validation, this segment monitors the degree of sparse categorical cross-entropy loss and accuracy metric fluctuations as epochs increase.

Accuracy Fluctuations Across Epochs in Training and Cross-Validation

Figure 4-8 depicts the degree of accuracy fluctuations as epochs increase in training and cross-validation when the CNN classifies patient breast cancer outcomes. The code is in Listing 4-14.

Listing 4-14. Charting Accuracy Fluctuations Across Epochs in Training and Cross-Validation

```
plt.plot(breast_cancer_convolutional_net_model_history.
history["accuracy"],
         color = "orange",
         marker = "o",
         label = "Training accuracy")
plt.plot(breast_cancer_convolutional_net_model_history.
history["val_accuracy"],
         color = "blue",
         marker = "o",
         label = "CV accuracy")
plt.xlabel("Epochs")
plt.ylabel("Accuracy")
plt.legend(loc = "best")
plt.show()
```

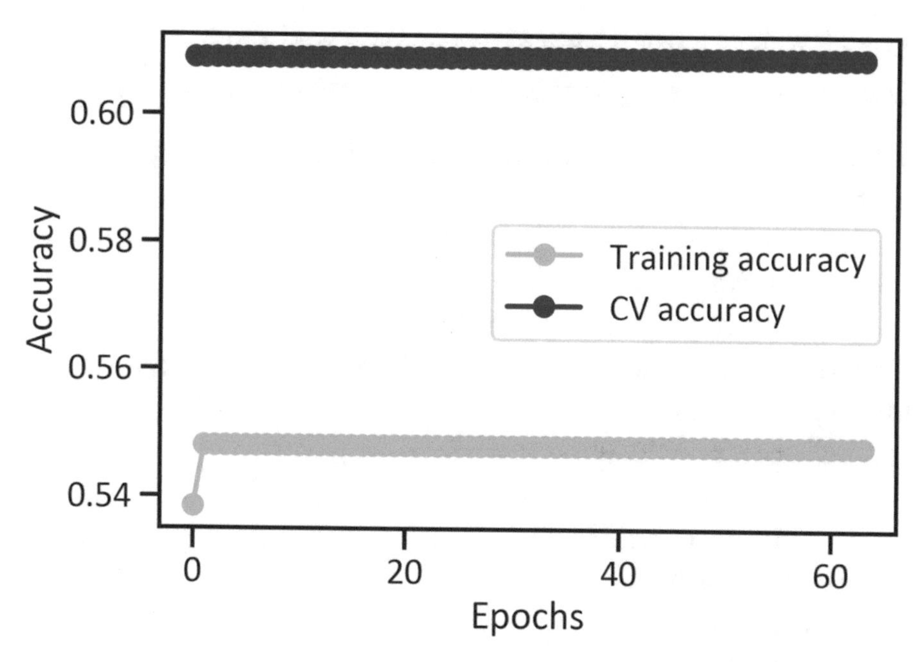

Figure 4-8. *Accuracy fluctuations across epochs in training and cross-validation*

Sparse Categorical Cross-Entropy Loss Fluctuations Across Epochs in Training and Cross-Validation

Figure 4-9 depicts the degree of sparse categorical cross-entropy loss fluctuations as epochs increase in training and cross-validation when the CNN classifies patient breast cancer outcomes. See the code in Listing 4-15.

Listing 4-15. Charting Sparse Categorical Cross-Entropy Loss Fluctuations Across Epochs in Training and Cross-Validation

```
plt.plot(breast_cancer_convolutional_net_model_history.
history["loss"],
         color = "orange",
         marker = "o",
         label = "Training loss")
plt.plot(breast_cancer_convolutional_net_model_history.
history["val_loss"],
         color = "blue",
         marker = "o",
         label = "CV loss")
plt.xlabel("Epochs")
plt.ylabel("Loss")
plt.legend(loc = "best")
plt.show()
```

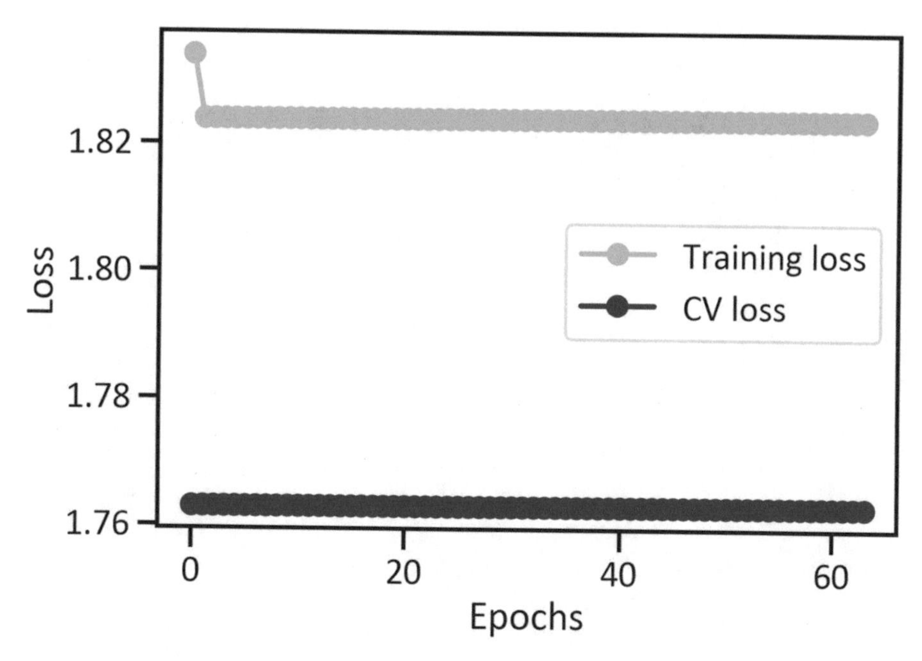

Figure 4-9. *Sparse categorical cross-entropy loss fluctuations across epochs in training and cross-validation*

Conclusion

This chapter executed convolutional neural networks to classify scan images of patients with and without cancer. It concluded by appraising both convolutional neural networks.

CHAPTER 5

Modeling Magnetic Resonance Imaging and X-Rays by Executing Artificial Neural Networks

This chapter acquaints you with the practical application of computer vision and artificial neural networks in neurology and radiology. Over the course of this chapter, you will execute convolutional neural networks (CNNs) for image classification. The initial network will model MRI scans to differentiate patients with and without a brain tumor, and the second network will model X-ray scans to differentiate patients with and without pneumonia. You'll learn an effective technique for appraising networks in medical image classification.

T. C. Nokeri, *Artificial Intelligence in Medical Sciences and Psychology*,
https://doi.org/10.1007/978-1-4842-8217-5_5

Brain Tumors

A brain tumor represents an abnormal growth of cells around the brain region. There are several types of brain tumors. Figure 5-1 depicts the most common brain tumors.

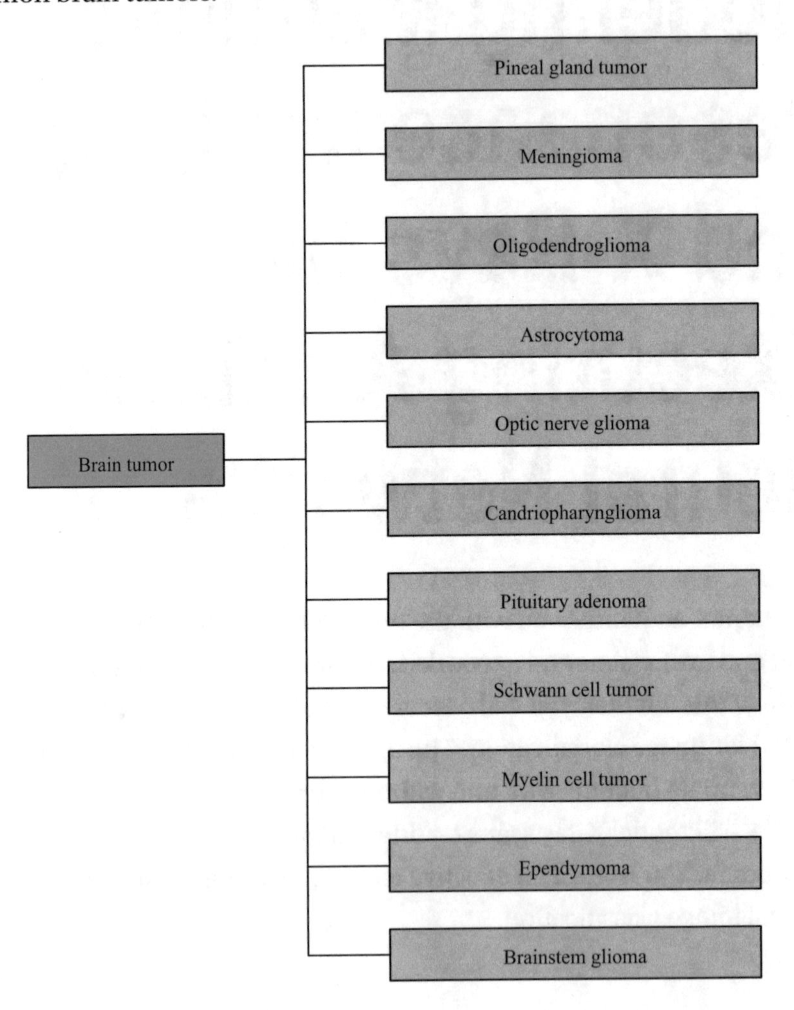

Figure 5-1. *Types of brain tumors*

MRI Procedure

The most prevalent method for uncovering brain tumors involves magnetic resonance imaging (MRI) scanning, which is performed by specialized healthcare professionals like neurologists and neurosurgeons, among others. Figure 5-2 provides a high-level overview of the MRI scanning procedure.

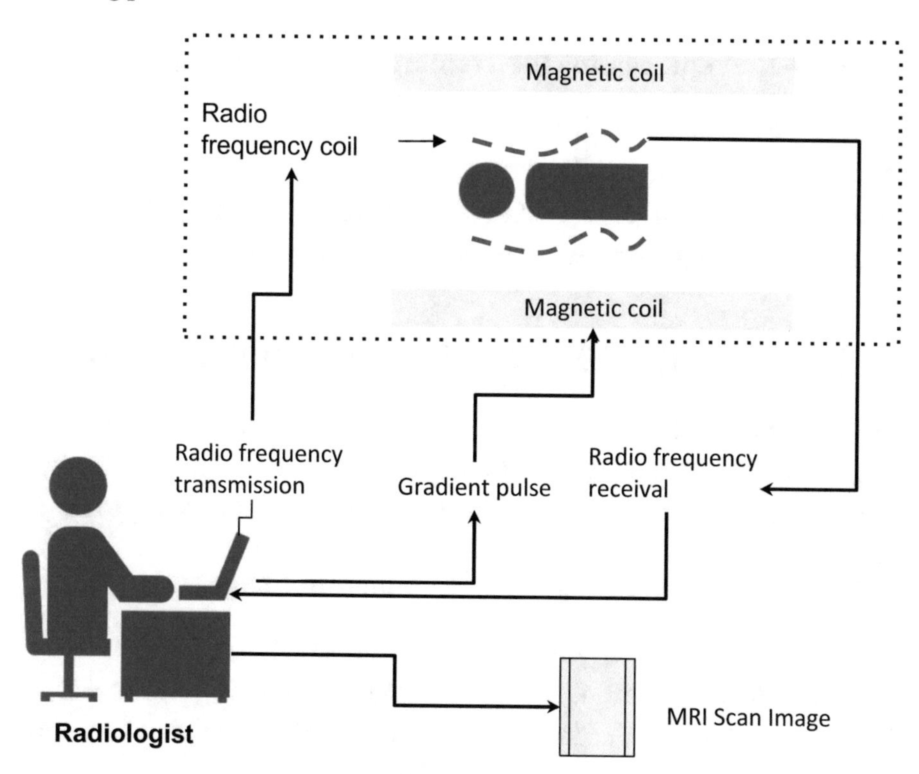

Figure 5-2. *Simple MRI procedure*

Figure 5-2 depicts a radiologist using a computer-based system that transmits radio frequencies received by radio frequency coils, in addition to gradient pulses received by a magnetic coil. In return, the computer-based system receives and processes feedback frequency to generate an MRI scan.

Preprocessing the Training MRI Image Data

Listing 5-1 preprocesses the images (for training) by executing the `image_dataset_from_directory()` method in the keras library. Equally, it outlines the `validation_split`, `subset`, `image_size`, and `batch_size`. Start by installing TensorFlow in your environment: `pip install tensorflow`.

Listing 5-1. Preprocessing the Training MRI Image Data

```
import tensorflow as tf
brain_tumor_training_data = image_dataset_from_directory(brain_
tumor_data, seed = 123, validation_split = 0.2, subset =
"training", image_size = (180, 180), batch_size = 16)
```

Preprocessing the Validation MRI Image Data

Listing 5-2 preprocesses the images (for validation) by executing the `image_dataset_from_directory()` method in the keras library. Equally, it outlines the `validation_split`, `subset`, `image_size`, and `batch_size`.

Listing 5-2. Preprocessing the Validation MRI Image Data

```
brain_tumor_validation_data = image_dataset_from_
directory(brain_tumor_data, seed = 123, validation_split = 0.2,
subset = "validation", image_size = (180, 180), batch_size = 16)
```

Generating the Training MRI Image Data

Listing 5-3 generates an image for training by executing the `ImageDataGenerator()` method and outlining the rescaled ratio. Following that, it implements the `flow_from_directory()` method and outlines the `target_size`, `class_mode`, and `batch_size`. Start by installing NumPy in your environment: `pip install numpy`.

Listing 5-3. Generating the Training MRI Image Data

```
import numpy as np
brain_tumor_training_data_categories = np.array(brain_tumor_
training_data.class_names)
brain_tumor_generated_image_data = ImageDataGenerator(rescale
= 1./255)
brain_tumor_generated_image_data_for_training = brain_tumor_
generated_image_data.flow_from_directory(brain_tumor_data,
target_size = (180, 180), class_mode = "categorical", shuffle =
True, batch_size = 16)
brain_tumor_images, brain_tumor_labels = next(iter(brain_tumor_
generated_image_data_for_training))
```

Tuning the Training MRI Image Data

This section develops a CNN. Initially, it prepares the image data in such a way that the network better identifies patterns by tuning them (see Listing 5-4).

Listing 5-4. Tuning the Training MRI Image Data

```
brain_tumor_experimental_tuning = tf.data.experimental.AUTOTUNE
```

```
brain_tumor_training_data = brain_tumor_training_data.
cache().shuffle(1000).prefetch(buffer_size = brain_tumor_
experimental_tuning)
brain_tumor_validation_data = brain_tumor_validation_data.
cache().prefetch(buffer_size = brain_tumor_experimental_tuning)
```

Executing the CNN to Classify MRI Image Data

Listing 5-5 executes the CNN on the image data.

Listing 5-5. Executing the CNN to Classify MRI Image Data

```
from tensorflow.keras import layers
from tensorflow.python.keras.layers import Dense, Flatten,
Conv2D, Dropout, MaxPooling2D
brain_tumor_convolutional_net_model = tf.keras.Sequential([
  tf.keras.layers.experimental.preprocessing.Rescaling(1./255),
  tf.keras.layers.Conv2D(16, 3, activation = "relu"),
  tf.keras.layers.MaxPooling2D(),
  tf.keras.layers.Conv2D(64, 3, activation="relu"),
  tf.keras.layers.MaxPooling2D(),
  tf.keras.layers.Conv2D(128, 3, activation="relu"),
  tf.keras.layers.MaxPooling2D(),
  tf.keras.layers.Flatten(),
  tf.keras.layers.Dense(255, activation="relu"),
  tf.keras.layers.Dense(3)
])
brain_tumor_convolutional_net_model.compile(optimizer = "adam",
loss = tf.keras.losses.SparseCategoricalCrossentropy(from_
logits = True), metrics = ["accuracy"])
```

```
brain_tumor_convolutional_net_model_history = brain_tumor_
convolutional_net_model.fit(brain_tumor_training_data,
validation_data = brain_tumor_validation_data, epochs = 64)
```

Considering the CNN's Performance

The CNN that classifies patient brain tumor outcomes comprises a set of 2D convolutional and MaxPooling 2D layers, a flatten layer, and two dense layers. All layers contain a `relu` activation function, consisting of a sparse cross-categorical cross-entropy loss function and an accuracy metric. Furthermore, it is trained across 64 epochs.

Accuracy Fluctuations Across Epochs in Training and Cross-Validation

Figure 5-3 depicts the degree of accuracy fluctuations as epochs increase in training and cross-validation when the CNN classifies patient brain tumor outcomes. See Listing 5-6 for the code.

Listing 5-6. Charting the CNN's Accuracy Score Across Epochs

```
plt.plot(brain_tumor_convolutional_net_model_history.
history["accuracy"],
         color = "orange",
         marker = "o",
         label = "Training accuracy")
plt.plot(brain_tumor_convolutional_net_model_history.
history["val_accuracy"],
         color = "blue",
         marker = "o",
         label = "CV accuracy")
```

```
plt.xlabel("Epochs")
plt.ylabel("Loss")
plt.legend(loc = "best")
plt.show()
```

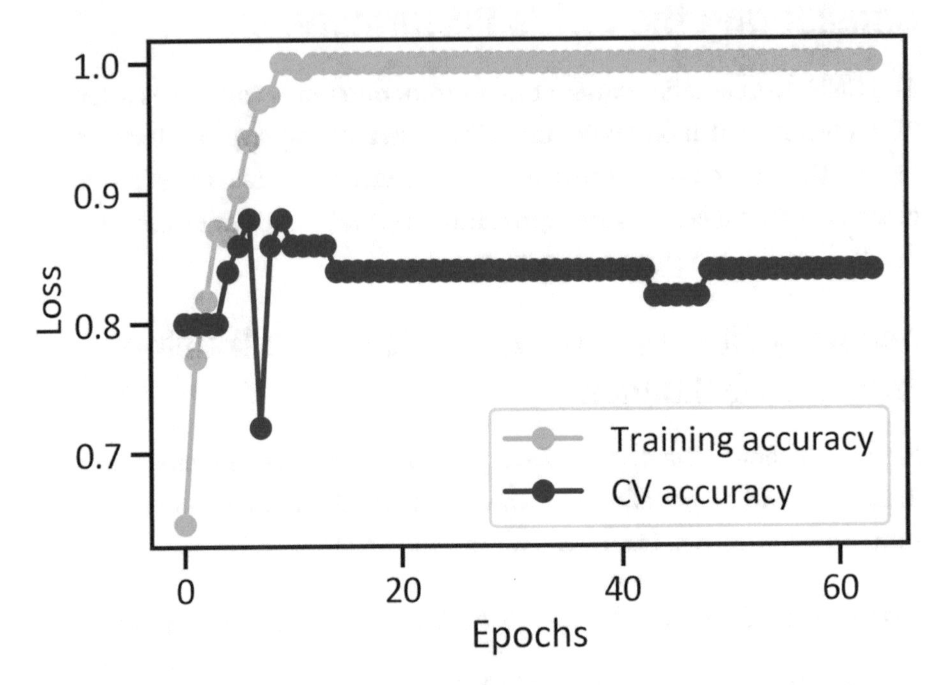

Figure 5-3. *Accuracy fluctuations across epochs in training and cross-validation*

Figure 5-3 depicts an accuracy increase in training and cross-validation when the CNN classifies patient brain tumor outcomes, with it being slightly lower in cross-validation across all epochs.

Sparse Categorical Cross-Entropy Loss Fluctuations Across Epochs in Training and Cross-Validation

Figure 5-4 depicts the degree of sparse categorical cross-entropy loss fluctuations as epochs increase in training and cross-validation when the CNN classifies patient brain tumor outcomes. See Listing 5-7 for the code.

Listing 5-7. Charting the CNN's Loss Across Epochs

```
plt.plot(brain_tumor_convolutional_net_model_history.
history["loss"],
         color = "orange",
         marker = "o",
         label = "Training loss")
plt.plot(brain_tumor_convolutional_net_model_history.
history["val_loss"],
         color = "blue",
         marker = "o",
         label = "CV loss")
plt.xlabel("Epochs")
plt.ylabel("Loss")
plt.legend(loc = "best")
plt.show()
```

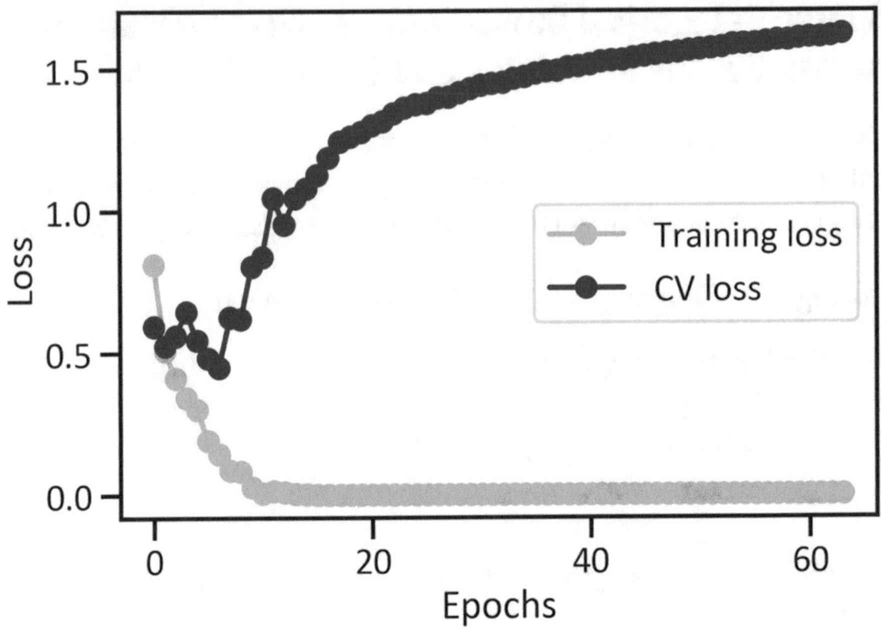

***Figure 5-4.** Sparse categorical cross-entropy loss fluctuations across epochs in training and cross-validation*

Figure 5-4 depicts a sparse categorical cross-entropy loss increase in training, but a decrease when the CNN classifies patient brain tumor outcomes.

Pneumonia

Pneumonia is a preventable infection in the lungs that derives from a bacteria or virus. To prevent it, one may opt for a vaccination. Typical symptoms of pneumonia include a persistent cough, prolonged fever, and breathing difficulties, among others. This disease is common among asthmatics, diabetic individuals, and smokers, among other risk factors. The branch of medical science that combats this disease is called pulmonology.

X-Ray Imaging Procedure

Figure 5-5 depicts the X-ray imaging procedure.

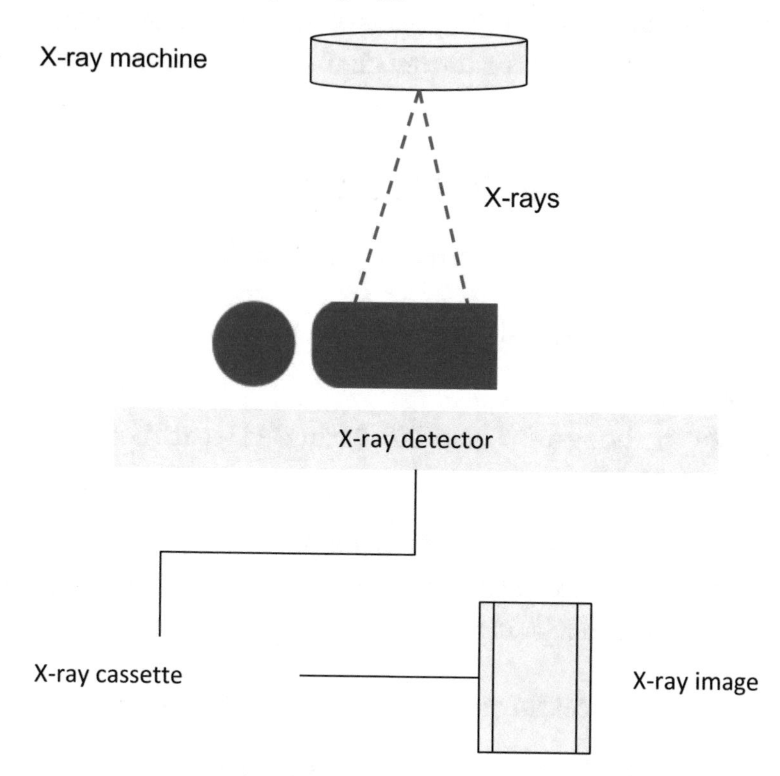

***Figure 5-5.** X-ray imaging procedure*

Figure 5-5 depicts the X-ray machine releasing radiation doses which pass through the body and are detected by the X-ray detector, which is then transmitted to the cassette and an image is generated.

Classifying X-Rays by Executing a CNN

This section covers computer vision techniques for filtering and the use of a CNN to classify chest X-ray scans of patients with and without pneumonia. The dataset was extracted here[1] and you can download from Kaggle[2].

Processing the X-Ray Image Data

Listing 5-8 collects the X-ray scan data for training the CNN.

Listing 5-8. Processing the Training Chest X-Ray Image Data

```
import pathlib
chest_x_ray_training_df = pathlib.Path(r"filepath\xray_
dataset\train")
```

Listing 5-9 preprocesses the images (for training) by executing the `image_dataset_from_directory()` method in the `keras` library. Equally, it outlines the `validation_split`, `subset`, `image_size`, and `batch_size`.

Listing 5-9. Preprocessing the Brain Tumor Training Image Data

```
chest_x_ray_training_data = image_dataset_from_
directory(chest_x_ray_training_df, seed = 123, validation_split
= 0.2, subset = "training",  image_size = (180, 180), batch_
size = 16)
```

[1] www.cell.com/cell/fulltext/S0092-8674(18)30154-5

[2] www.kaggle.com/paultimothymooney/chest-xray-pneumonia

Generating the Training Chest X-Ray Image Data

Listing 5-10 generates an image for training by executing the ImageDataGenerator() method and outlining the rescaled ratio. Following that, it implements the flow_from_directory() method and outlines the target_size, class_mode, and batch_size.

Listing 5-10. Generating the Training Chest X-Ray Image Data

```
chest_x_ray_training_data_categories = np.array(chest_x_ray_
training_data.class_names)
chest_x_ray_generated_image_data = ImageDataGenerator(rescale
= 1./255)
chest_x_ray_generated_image_data_for_training = chest_x_
ray_generated_image_data.flow_from_directory(chest_x_
ray_training_df, target_size = (180, 180), class_mode =
"categorical",  shuffle = True,  batch_size = 16)
chest_x_ray_training_images, chest_x_ray_training_labels =
next(iter(chest_x_ray_generated_image_data_for_training))
```

Preprocessing the Validation Chest X-Ray Image Data

Listing 5-11 collects the X-ray image data for validating the CNN.

Listing 5-11. Processing the Validation Chest X-Ray Image Data

```
chest_x_ray_validation_df = r"filepath\xray_dataset\test"
```

Listing 5-12 preprocesses the images (for training) by executing the image_dataset_from_directory() method in the keras library. Equally, it outlines the validation_split, subset, image_size, and batch_size.

Listing 5-12. Preprocessing the Validation Chest X-Ray Image Data

```
chest_x_ray_validation_data = image_dataset_from_
directory(chest_x_ray_validation_df, seed = 123, validation_
split = 0.2, subset = "validation",  image_size = (180, 180),
batch_size = 16)
```

Generating the Validation Chest X-Ray Image Data

Listing 5-13 generates an image for training by executing the `ImageDataGenerator()` method and outlining the rescaled ratio. Following that, it implements the `flow_from_directory()` method and outlines the `target_size`, `class_mode`, and `batch_size`.

Listing 5-13. Generating the Validation Chest X-Ray Image Data

```
chest_x_ray_validation_data_categories = np.array(chest_x_ray_
validation_data.class_names)
chest_x_ray_generated_image_data = ImageDataGenerator(rescale
= 1./255)
chest_x_ray_generated_image_data_for_validation = chest_x_
ray_generated_image_data.flow_from_directory(chest_x_ray_
validation_df, target_size = (180, 180), class_mode =
"categorical", shuffle = True, batch_size = 16)
chest_x_ray_validation_images, chest_x_ray_validation_labels =
next(iter(chest_x_ray_generated_image_data_for_validation))
```

Tuning the Training Chest X-Ray Image Data

This section develops a CNN. Initially, it prepares the image data in such a way that the network better identifies patterns by tuning them (see Listing 5-14).

Listing 5-14. Tuning the Training Chest X-Ray Image Data

```
chest_x_ray_experimental_tuning = tf.data.experimental.AUTOTUNE
chest_x_ray_training_data = chest_x_ray_training_data.
cache().shuffle(1000).prefetch(buffer_size = chest_x_ray_
experimental_tuning)
chest_x_ray_validation_data = chest_x_ray_validation_data.
cache().prefetch(buffer_size = chest_x_ray_experimental_tuning)
```

Executing the CNN to Classify Chest X-Ray Image Data

The architectural structure of the CNN that classifies patient pneumonia outcomes is like that of the CNN above that classifies patient brain tumor outcomes. It also comprises a set of 2D convolutional and MaxPooling 2D layers, a flatten layer, and two dense layers. All layers contain a `relu` activation function, consisting of a sparse cross-categorical cross-entropy loss function, and an accuracy metric. Furthermore, it is trained across 64 epochs. See the code in Listing 5-15.

Listing 5-15. Executing the CNN to Classify Chest X-Ray Image Data

```
chest_x_ray_convolutional_net_model = tf.keras.Sequential([
  tf.keras.layers.experimental.preprocessing.Rescaling(1./255),
  tf.keras.layers.Conv2D(16, 3, activation = "relu"),
  tf.keras.layers.MaxPooling2D(),
  tf.keras.layers.Conv2D(64, 3, activation="relu"),
  tf.keras.layers.MaxPooling2D(),
  tf.keras.layers.Conv2D(128, 3, activation="relu"),
  tf.keras.layers.MaxPooling2D(),
  tf.keras.layers.Flatten(),
```

```
  tf.keras.layers.Dense(255, activation="relu"),
  tf.keras.layers.Dense(3)
])
chest_x_ray_convolutional_net_model.compile(optimizer = "adam",
loss = tf.keras.losses.SparseCategoricalCrossentropy(from_
logits = True), metrics = ["accuracy"])
chest_x_ray_convolutional_net_model_history = chest_x_ray_
convolutional_net_model.fit(chest_x_ray_training_data,
validation_data = chest_x_ray_validation_data, epochs = 64)
```

Considering the CNN's Performance

To determine how well the CNN classifies patient pneumonia outcomes in training and cross-validation, this segment monitors the degree of sparse categorical cross-entropy loss and accuracy metric fluctuations as epochs increase.

Accuracy Fluctuations Across Epochs in Training and Cross-Validation

Figure 5-6 depicts the degree of accuracy fluctuations as epochs increase in training and cross-validation when the CNN classifies patient pneumonia outcomes. See the code in Listing 5-16.

Listing 5-16. Charting Accuracy Fluctuations Across Epochs in Training and Cross-Validation

```
plt.plot(chest_x_ray_convolutional_net_model_history.
history["accuracy"],
         color = "orange",
         marker = "o",
         label = "Training accuracy")
```

```
plt.plot(chest_x_ray_convolutional_net_model_history.
history["val_accuracy"],
         color = "blue",
         marker = "o",
         label = "CV accuracy")
plt.xlabel("Epochs")
plt.ylabel("Loss")
plt.legend(loc = "best")
plt.show()
```

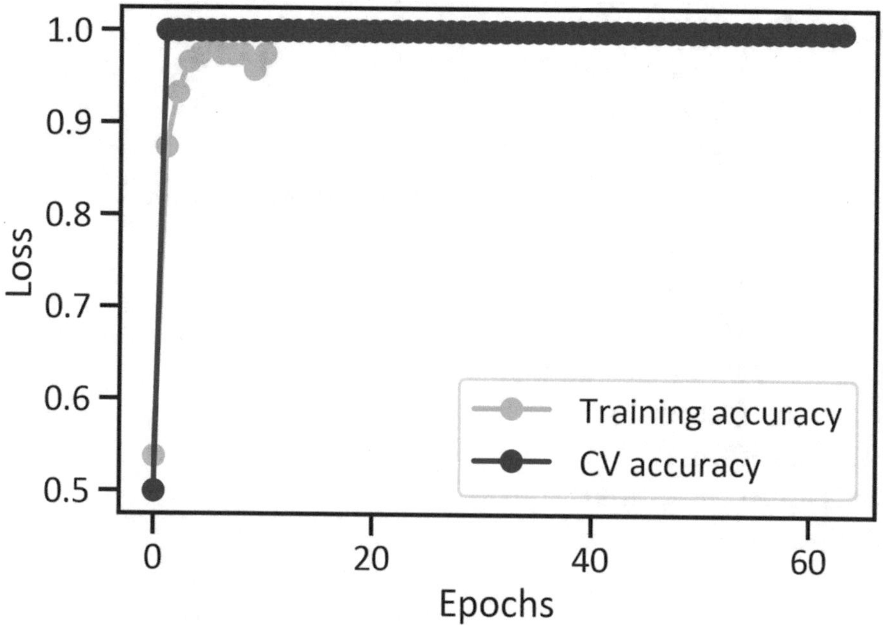

Figure 5-6. *Accuracy fluctuations across epochs in training and cross-validation*

Figure 5-6 depicts a sudden accuracy expansion in training and cross-validation when the CNN classifies patient pneumonia outcomes, with it being slightly lower in cross-validation.

Sparse Categorical Cross-Entropy Loss Fluctuations Across Epochs in Training and Cross-Validation

Figure 5-7 depicts the degree of sparse categorical cross-entropy loss fluctuations as epochs increase in training and cross-validation when the CNN classifies patient pneumonia outcomes. See the code in Listing 5-17.

Listing 5-17. Charting Sparse Categorical Cross-Entropy Loss Fluctuations Across Epochs in Training and Cross-Validation

```
plt.plot(chest_x_ray_convolutional_net_model_history.
history["loss"],
         color = "orange",
         marker = "o",
         label = "Training loss")
plt.plot(chest_x_ray_convolutional_net_model_history.
history["val_loss"],
         color = "blue",
         marker = "o",
         label = "CV loss")
plt.xlabel("Epochs")
plt.ylabel("Loss")
plt.legend(loc = "best")
plt.show()
```

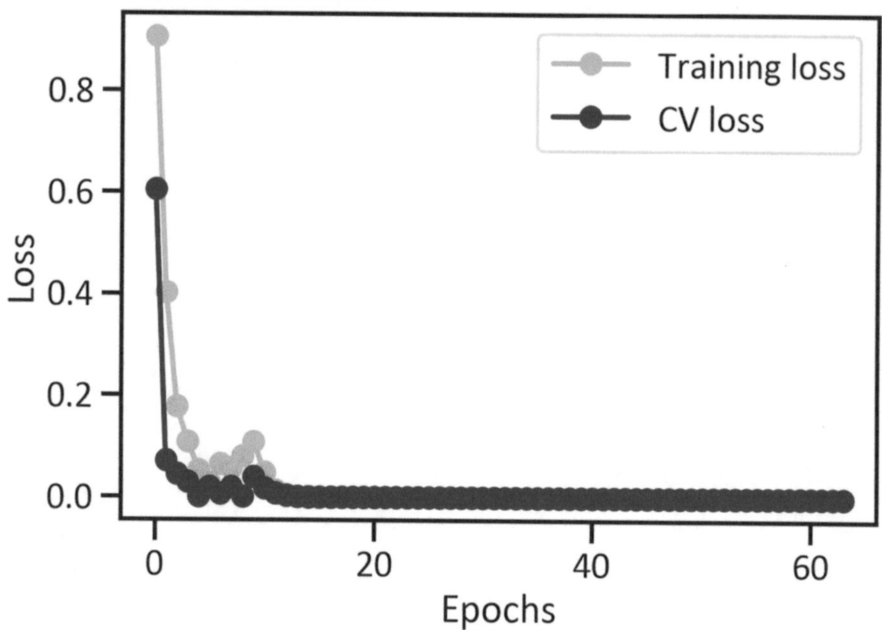

Figure 5-7. *Sparse categorical cross-entropy loss fluctuations across epochs in training and cross-validation*

Figure 5-7 depicts the degree of sparse categorical cross-entropy loss fluctuations as epochs increase in training and cross-validation when the CNN classifies patient pneumonia outcomes.

Conclusion

This chapter introduced two standard libraries for computer vision (i.e., `OpenCV`) and artificial neural network development (i.e., `TensorFlow/Keras`). The subsequent chapter extends this chapter with a focus on breast and skin cancer realization and segmentation.

CHAPTER 6

A Case for COVID-19 CT Scan Segmentation

This chapter presents an approach for carrying out convolutional neural networks to model chest CT scan images and differentiate between patients with and without COVID-19. You can download the dataset from Kaggle[1]; the initial dataset comes from here[2].

A Simple Computer Tomography Scan Procedure

A nasal swab test is used in COVID-19 testing; it involves inserting a swab inside the nasal passage to extract cells. Moreover, radiologists can perform a computer tomography (CT) scan to get an image that helps determine whether an individual is COVID-19 positive. Figure 6-1 depicts a simple CT scan procedure.

[1] www.kaggle.com/khoongweihao/covid19-xray-dataset-train-test-sets
[2] https://github.com/ieee8023/covid-chestxray-dataset

T. C. Nokeri, *Artificial Intelligence in Medical Sciences and Psychology*,
https://doi.org/10.1007/978-1-4842-8217-5_6

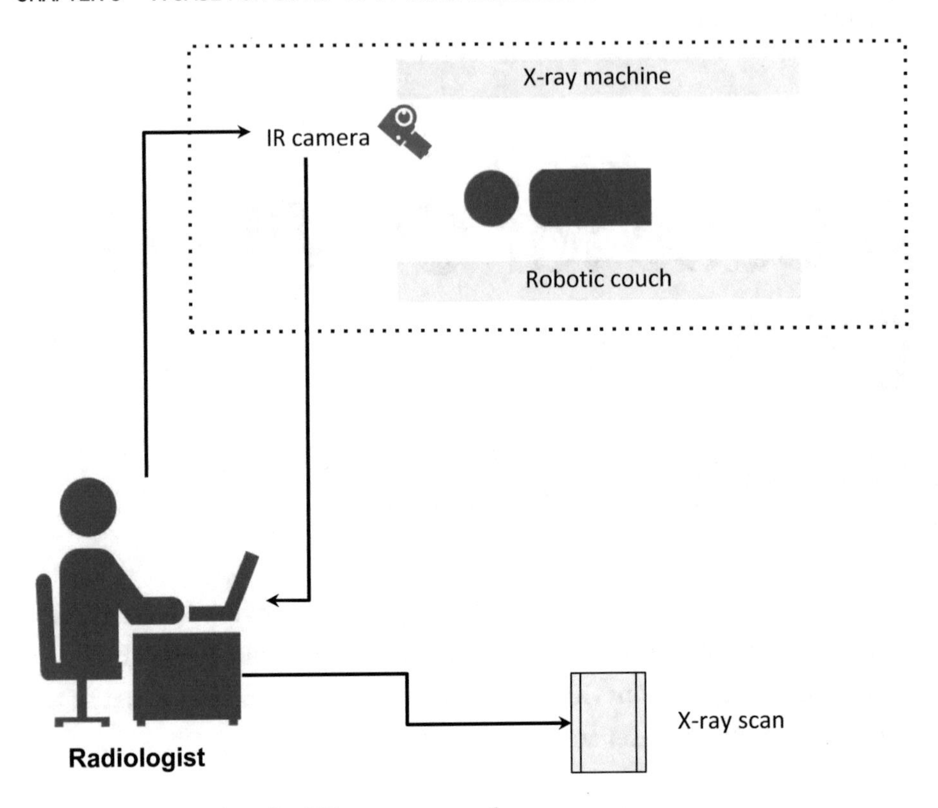

Figure 6-1. *A simple CT scan procedure*

Figure 6-1 depicts a radiologist using a computer-based system to operate a non-enclosed CT machine, which releases a few radiation doses while capturing the body at different angles using an infrared camera.

Preprocessing the Training COVID-19 Data

Listing 6-1 preprocesses the images (for training) by carrying out the `image_dataset_from_directory()` method from the `keras` library. Equally, it outlines the `validation_split`, `subset`, `image_size`, and `batch_size`. Start by installing Keras in your environment: `pip install keras`. In addition, install NumPy in your environment: `pip install numpy`.

Listing 6-1. Preprocessing the Training COVID-19 CT Scan Data

```
import keras
import numpy as np
import pathlib
covid_19_data = pathlib.Path(r"filepath\cov_19_ct_scans")
covid_19_training_data = image_dataset_from_directory(covid_19_
data,seed = 123, validation_split = 0.2, subset =
"training",  image_size = (180, 180), batch_size = 16)
```

Preprocessing the Validation COVID-19 CT Scan Data

Listing 6-2 preprocesses the images (for validation) by carrying out the `image_dataset_from_directory()` method from the keras library. Equally, it outlines the `validation_split`, `subset`, `image_size`, and `batch_size`.

Listing 6-2. Preprocessing the Validation COVID-19 CT Scan Data

```
covid_19_validation_data = image_dataset_from_
directory(covid_19_data, seed = 123, validation_split =
0.2, subset = "validation", image_size = (180, 180), batch_
size = 16)
```

Generating the Training COVID-19 CT Scan Data

Listing 6-3 generates an image for training by carrying out the `ImageDataGenerator()` method and outlining the rescaled ratio. Following that, it executes the `flow_from_directory()` method and outlines the `target_size`, `class_mode`, and `batch_size`.

Listing 6-3. Generating the Training COVID-19 CT Scan Data

```
import numpy as np
covid_19_training_data_categories = np.array(covid_19_training_
data.class_names)
covid_19_generated_image_data = keras.preprocessing.image.
ImageDataGenerator(rescale = 1./255)
covid_19_generated_image_data_for_training = covid_19_
generated_image_data.flow_from_directory(covid_19_
data,  target_size = (180, 180),  class_mode =
"categorical",  shuffle = True, batch_size = 16)
covid_19_images, covid_19_labels = next(iter(covid_19_
generated_image_data_for_training))
```

Tuning the Training COVID-19 CT Scan Data

This section develops a convolutional neural network. Initially, it prepares the image data in such a way that the network better identifies patterns by tuning them (see Listing 6-4).

Listing 6-4. Tuning the Training COVID-19 CT Scan Data

```
covid_19_experimental_tuning = tf.data.experimental.AUTOTUNE
covid_19_training_data = covid_19_training_data.
cache().shuffle(1000).prefetch(buffer_size = covid_19_
experimental_tuning)
covid_19_validation_data = covid_19_validation_data.cache().
prefetch(buffer_size = covid_19_experimental_tuning)
```

Carrying Out the CNN to Classify COVID-19 CT Scan Data

Listing 6-5 carries out the convolutional neural network on the image data. Start by installing TensorFlow in your environment: `pip install tensorflow`. The architectural structure of the CNN constructed in this chapter is similar to those constructed in the preceding chapter to classify patient brain tumor and pneumonia outcomes. It also comprises a set of 2D convolutional and MaxPooling 2D layers, a flatten layer, and two dense layers; all layers contain a `relu` activation function, consisting of a sparse cross-categorical cross-entropy loss function, and an accuracy metric. Furthermore, it is trained across 64 epochs.

Listing 6-5. Carrying Out the CNN to Classify COVID-19 CT Scan Data

```
from tensorflow.keras import layers
from tensorflow.python.keras.layers import Dense, Flatten,
Conv2D, Dropout, MaxPooling2D
covid_19_convolutional_net_model = tf.keras.Sequential([
  tf.keras.layers.experimental.preprocessing.Rescaling(1./255),
  tf.keras.layers.Conv2D(16, 3, activation = "relu"),
  tf.keras.layers.MaxPooling2D(),
  tf.keras.layers.Conv2D(64, 3, activation="relu"),
  tf.keras.layers.MaxPooling2D(),
  tf.keras.layers.Conv2D(128, 3, activation="relu"),
  tf.keras.layers.MaxPooling2D(),
  tf.keras.layers.Flatten(),
  tf.keras.layers.Dense(255, activation="relu"),
  tf.keras.layers.Dense(3)
])
```

```
covid_19_convolutional_net_model.compile(optimizer =
"adam",   = tf.keras.losses.SparseCategoricalCrossentropy
(from_logits = True), metrics = ["accuracy"])
covid_19_convolutional_net_model_history = covid_19_
convolutional_net_model.fit(covid_19_training_data, validation_
data = covid_19_validation_data, epochs = 64)
```

Considering the CNN's Performance

To determine how well the CNN classifies patient Covid-19 outcomes in training and cross-validation, this segment monitors the degree of sparse categorical cross-entropy loss and accuracy metric fluctuations as epochs increase.

Accuracy Fluctuations Across Epochs in Training and Cross-Validation

Figure 6-2 depicts the degree of accuracy fluctuations as epochs increase in training and cross-validation when the CNN classifies patient Covid-19 outcomes. See the code in Listing 6-6.

Listing 6-6. Charting Accuracy Fluctuations Across Epochs in Training and Cross-Validation

```
plt.plot(covid_19_convolutional_net_model_history.
history["accuracy"],
         color = "orange",
         marker = "o",
         label = "Training accuracy")
plt.plot(covid_19_convolutional_net_model_history.history["val_
accuracy"],
```

```
        color = "blue",
        marker = "o",
        label = "CV accuracy")
plt.xlabel("Epochs")
plt.ylabel("Accuracy")
plt.legend(loc = "best")
plt.show()
```

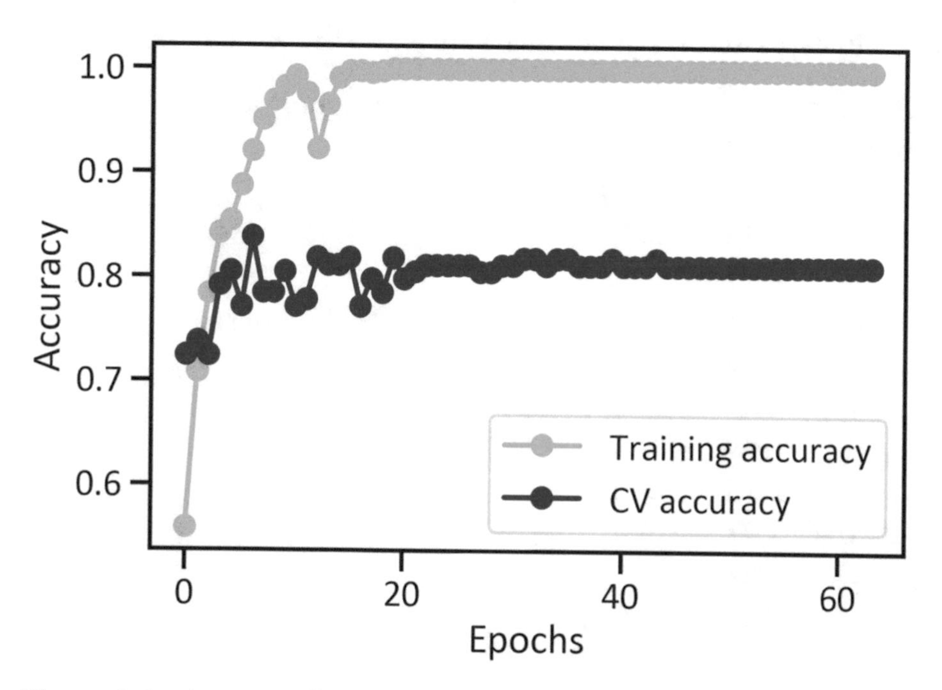

***Figure 6-2.** Accuracy fluctuations across epochs in training and cross-validation*

Figure 6-2 depicts accuracy increase as epochs increase in training and cross-validation, with it being higher during training than cross-validation.

Sparse Categorical Cross-Entropy Loss Fluctuations Across Epochs in Training and Cross-Validation

Figure 6-3 depicts the degree of sparse categorical cross-entropy loss fluctuations as epochs increase in training and cross-validation when the CNN classifies patient Covid-19 outcomes. See Listing 6-7.

Listing 6-7. Charting Sparse Categorical Cross-Entropy Loss Fluctuations Across Epochs in Training and Cross-Validation

```
plt.plot(covid_19_convolutional_net_model_history.
history["loss"],
         color = "orange",
         marker = "o",
         label = "Training loss")
plt.plot(covid_19_convolutional_net_model_history.
history["val_loss"],
         color = "blue",
         marker = "o",
         label = "CV loss")
plt.xlabel("Epochs")
plt.ylabel("Loss")
plt.legend(loc = "best")
plt.show()
```

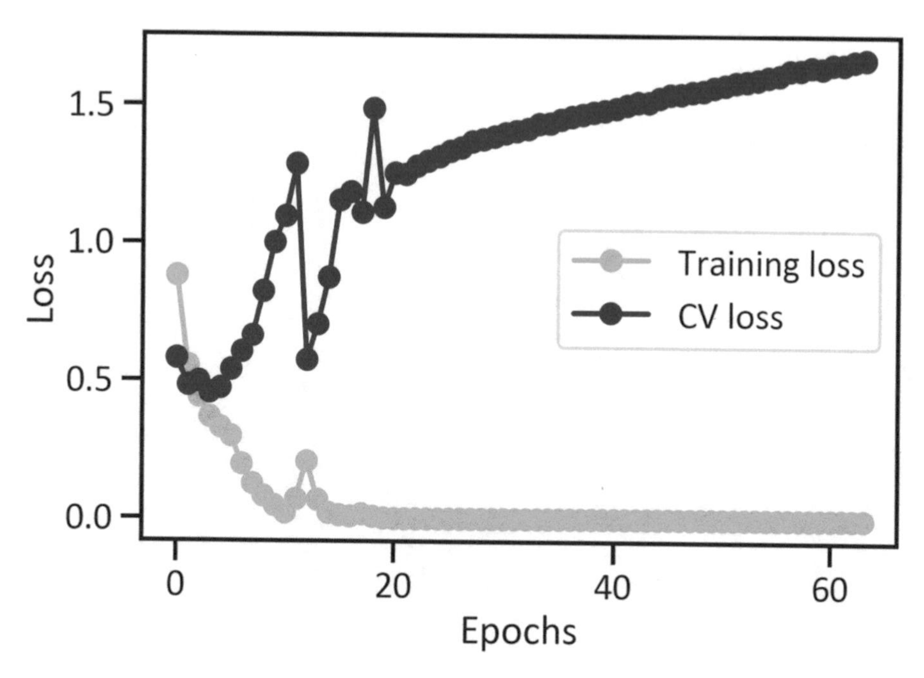

Figure 6-3. *Sparse categorical cross-entropy loss fluctuations across epochs in training and cross-validation*

Figure 6-3 depicts in training, the sparse categorical cross-entropy loss slightly declines in the first two epochs, followed by an upward trend until the 10th epoch, but the trend is erased by the 11th epoch and further expands until the 64th epoch. Meanwhile, the sparse categorical cross-entropy loss slowly deteriorates as epochs increase.

Conclusion

This chapter introduced you to the chest CT scan image classification for COVID-19 by a convolutional neural network via an approach similar to the two prior chapters.

CHAPTER 7

Modeling Clinical Trial Data

This chapter explains the prime essentials of the most widespread method for adequately investigating data from a clinical trial, recognized as a survival method. It also explains the Nelson-Aalen additive model. To begin, you'll explore the method. Then you'll go through exploratory analysis and then correlation analysis by carrying out the Pearson correlation method. Following that, you'll learn about the survival table and fit the model. Finally, you'll learn about the profile table and confidence interval and then you'll reproduce the cumulative and baseline hazard.

Clinical Trials

A clinical trial is an investigation that studies changes in a group of subjects upon intervention in a controlled environment over an extended period. This distinct research approach is widespread in medicine and psychology. It's considered more reliable than other investigations (i.e., cross-sectional investigation) since it occurs in a controlled environment. Besides that, it typically involves a controlled group, which is a set of subjects.

T. C. Nokeri, *Artificial Intelligence in Medical Sciences and Psychology*,
https://doi.org/10.1007/978-1-4842-8217-5_7

An Overview of Survival Analysis

When dealing with clinical trial data, we usually execute survival methods. These methods are a suitable candidate for data that comprises time components with missing values, equally recognized as censored data.

Table 7-1 outlines two forms of groups of research subjects that participate in a clinical trial.

Table 7-1. *Groups in a Clinical Trial*

Two forms of survival analysis families	Description
Survival classifiers	Making up survival methods that solve time-event problem, where there is censored data (missing data due to subjects in a clinical trial leaving the trial prior to conclusion) that comprise a dependent feature or an event column with classes. These are probabilistic methods. For instance, we apply survival classifiers when we want to determine the likelihood of a patient surviving an illness (or not dying from a certain illness under examination), considering time and censored events.
Survival regressors	Making survival methods that solve censored data comprise a dependent feature (or event column) that is continuous. Similar to survival classifiers, these methods are used for time-event problem solving with censored data, but they apply linear-based frameworks (using the hazard function) to address these problems. In addition, the outcome derives from a combination of covariates.

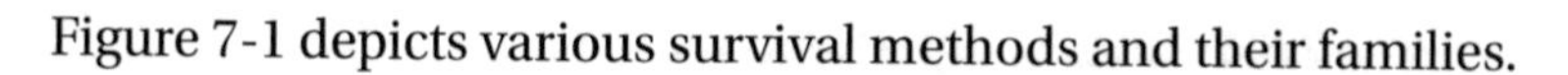

Figure 7-1 depicts various survival methods and their families.

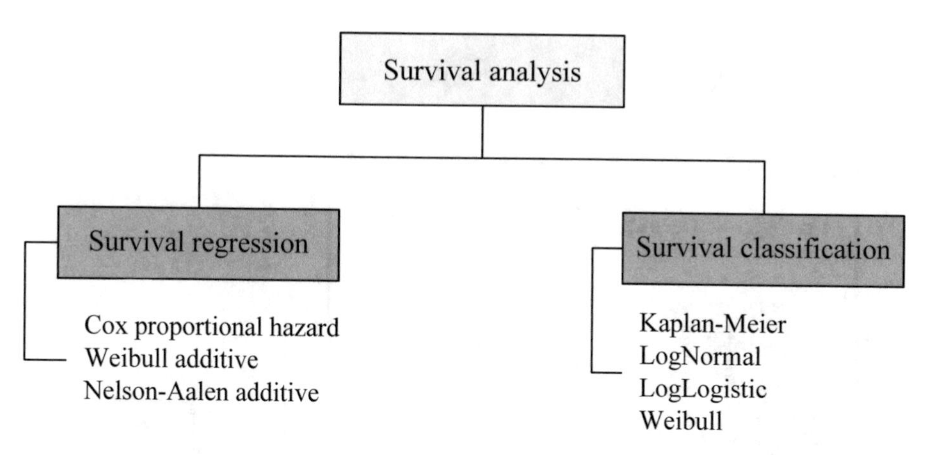

Figure 7-1. *Survival analysis families*

In the medical context, let's assume that there a pharmaceutical firm has discovered a new drug. Before the firm takes the drug into the market, it ought to comply with the rules and regulations that administer drugs. More importantly, the firm ought to ensure that the drug is effective.

The method requires you to assign research subjects to two groups (see Table 7-2).

Table 7-2. *Groups in a Clinical Trial*

Type of group	Description
Uncontrolled group	A group that is affected by the intervention (i.e., patients who consume the drug)
Controlled group	A group of research subjects that are not impacted by an intervention (i.e., patients who do not consume the drug)

See Figure 7-2 to further comprehend the example.

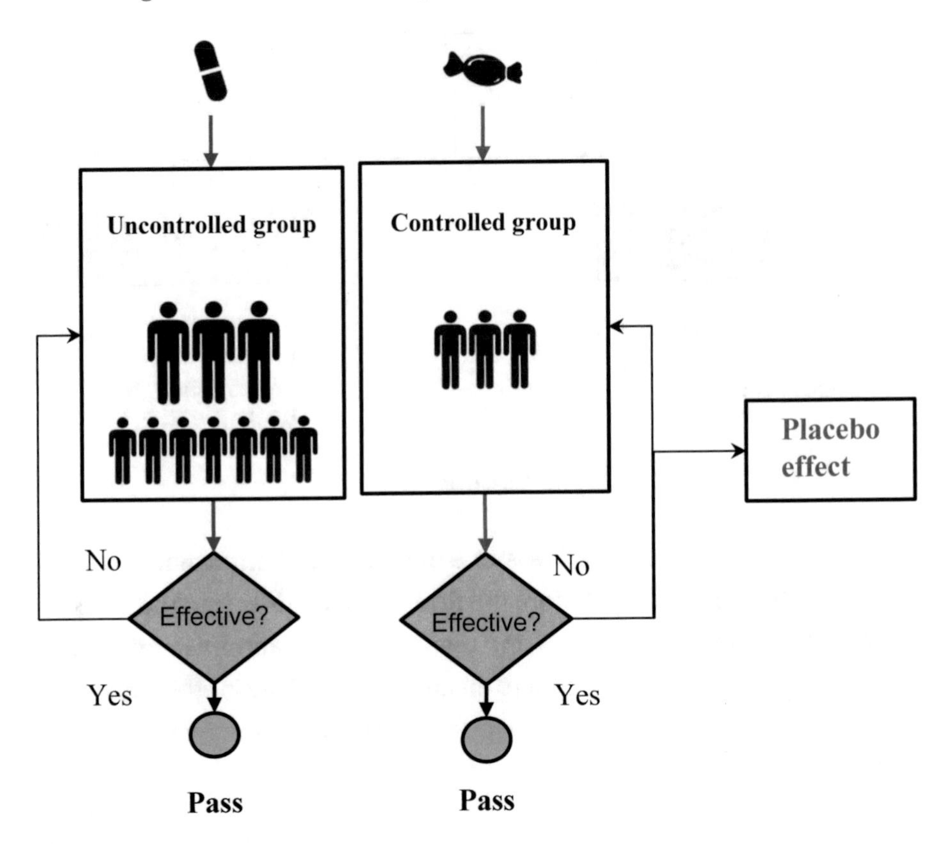

Figure 7-2. *Clinical trial setup*

Figure 7-2 depicts the setup of a clinical trial.

Clinical research scientists investigate the effects of the drug on the controlled group. In addition, they discern the placebo effect (a condition whereby the uncontrolled group exhibits effects similar to those of the controlled, but not as a result of consuming the drug). The presence of the placebo effect suggests the ineffectiveness of the drug.

Context of the Chapter

This chapter executes a survival regression method known as the Nelson-Aalen additive method. The dataset contains records relating to patients who underwent an operation to address breast cancer from 1958 to 1970 (University of Chicago's Billings Hospital, 1999). Download the dataset here[1].

Exploring the Nelson-Aalen Additive Model

The Nelson-Aalen Additive method is a prevalent survival regression method. It realizes the probability of survival at a specified time (t) by carrying out the hazard function. Equation 7-1 defines the Nelson-Aalen additive hazard function:

$$\tilde{H}_t = \sum_{t \leq 1} \frac{d_i}{n_i} \qquad \text{(Equation 7-1)}$$

where d_i signifies the events across time (t) and n_i signifies the number of subjects at risk.

The method will predict the likelihood of a patient's survival status upon an intervention at a certain time (the `Year_of_operation`).

Descriptive Analysis

Before you execute the Nelson-Aalen additive model, let's study the distribution of the data and I'll explain it. Initially, it counts values.

[1] `www.kaggle.com/gilsousa/habermans-survival-data-set`

Listing 7-1 collects and renames columns of the censored data accordingly. Start by installing pandas in your environment: `pip install pandas`.

Listing 7-1. Collecting Censored Data

```
import pandas as pd
survival_censored_data = pd.read_csv(r"filepath\haberman_
survival_data\haberman_survival_data.csv")
survival_censored_data.columns = ["Age", "Year_of_operation",
"Axillary_lymph_nodes ", "Event"]
```

Listing 7-2 realizes missing values in the survival data (see Figure 7-3). Start by installing seaborn (`pip install seaborn`) and Matplotlib (`pip install regex`) in your environment.

Listing 7-2. Realizing Missing Values

```
import seaborn as sns
import matplotlib.pyplot as plt
%matplotlib inline
sns.heatmap(survival_censored_data.isnull())
plt.show()
```

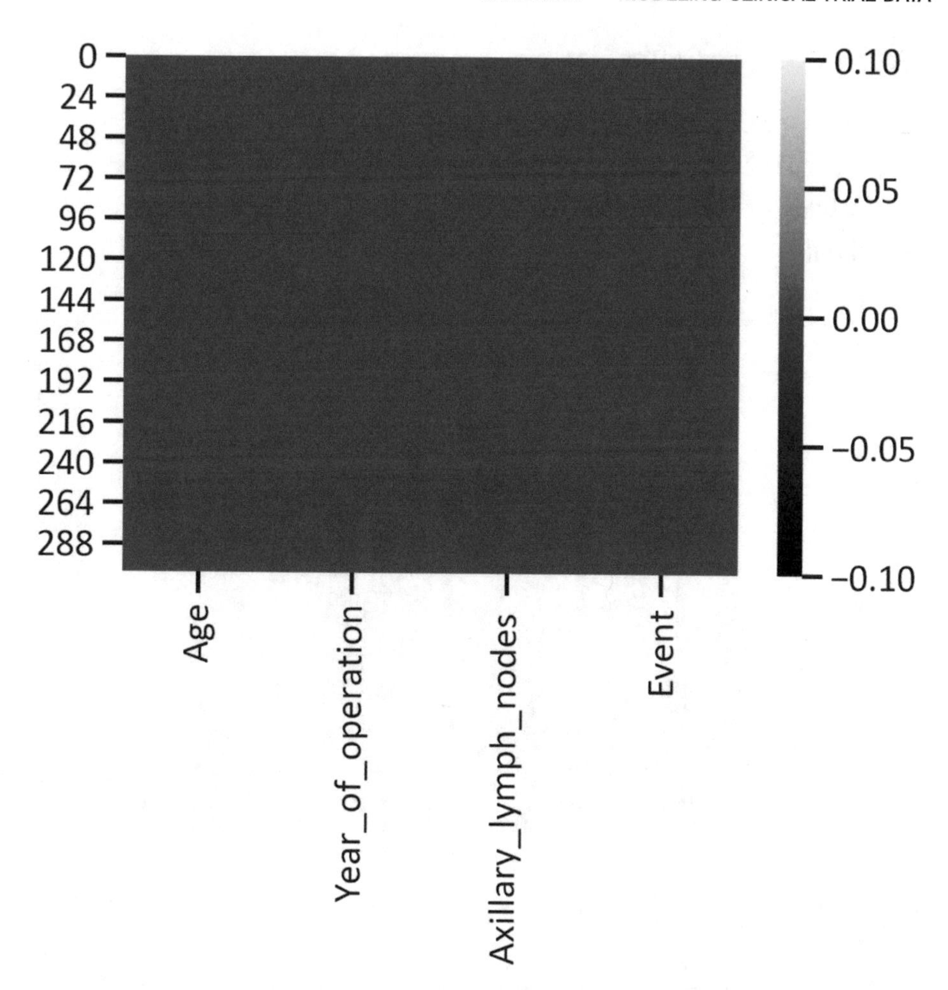

Figure 7-3. *Heatmap for null values*

Figure 7-3 signals the absence of null values in the survival data.

Listing 7-3 depicts a count plot of the age feature in the survival data (see Figure 7-4). It exhibits the feature (age) on the `x-axis` and the number of occurrences on the `y-axis`.

Listing 7-3. Age Count Plot

```
fig, ax = plt.subplots(figsize=(12, 7))
sns.countplot(survival_censored_data["Age"], ax = ax)
plt.xticks (rotation = 90)
plt.show()
```

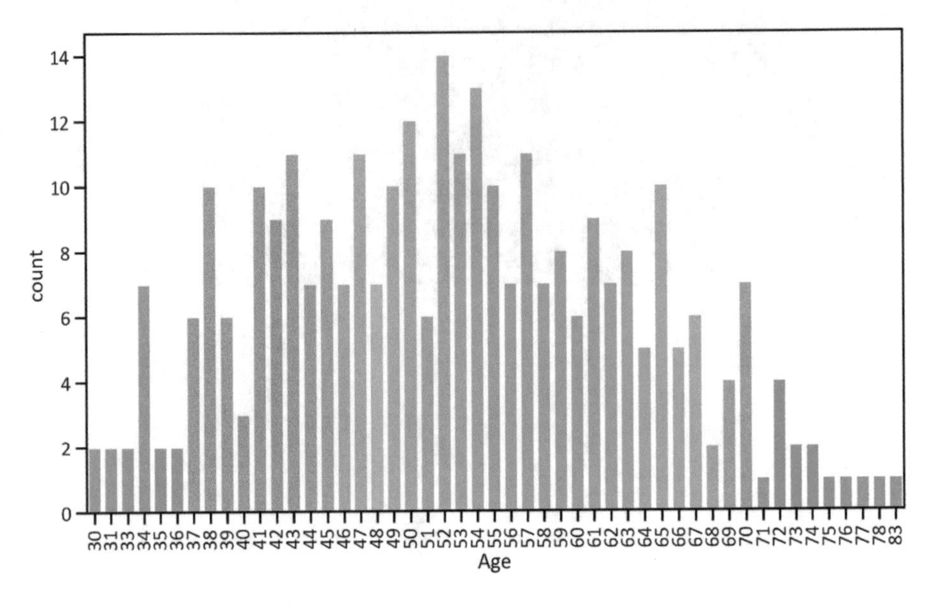

Figure 7-4. *Age count plot*

Figure 7-4 signals that patients at age 52 appear the most in the data, followed by patients at age 54. Patients at ages 71 and 75-83 appear the least in the data (they all appear once).

Listing 7-4 depicts a count plot of the `Year_of_operation` feature in the survival data (see Figure 7-5).

Listing 7-4. Year of Operation Count Plot

```
fig, ax = plt.subplots(figsize=(12, 7))
sns.countplot(survival_censored_data["Year_of_operation"],
ax = ax)
plt.xticks (rotation = 90)
plt.show()
```

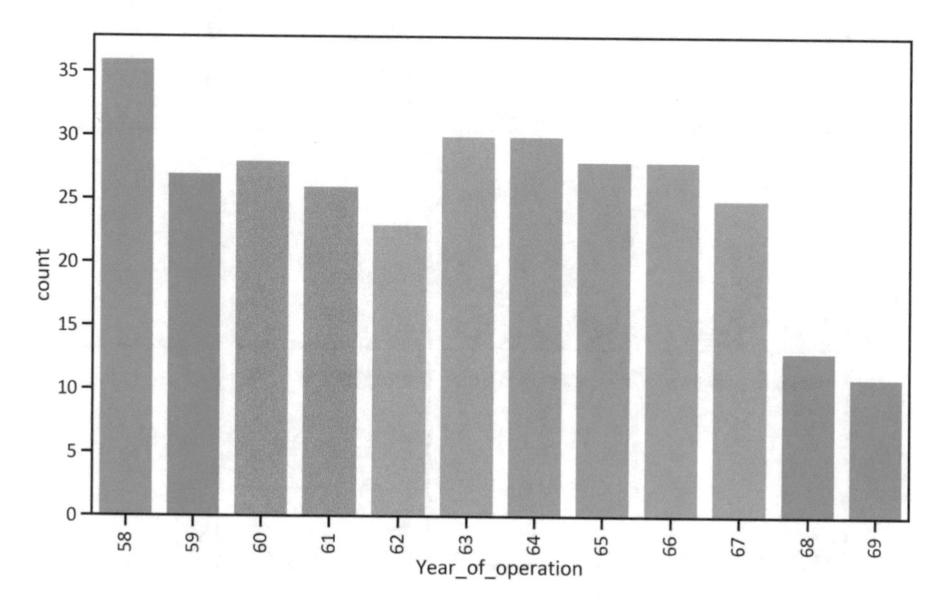

Figure 7-5. *Year_of_operation count plot*

Figure 7-5 shows that most operations were conducted in 1958, followed by 1963-1964. In 1969, there were 10 operations, which was the least of all.

Listing 7-5 depicts the survival data pair plot to identify the distribution of the data, including the relationships among the features (see Figure 7-6).

Listing 7-5. Survival Data Pair Plot

```
sns.pairplot(survival_censored_data)
plt.show()
```

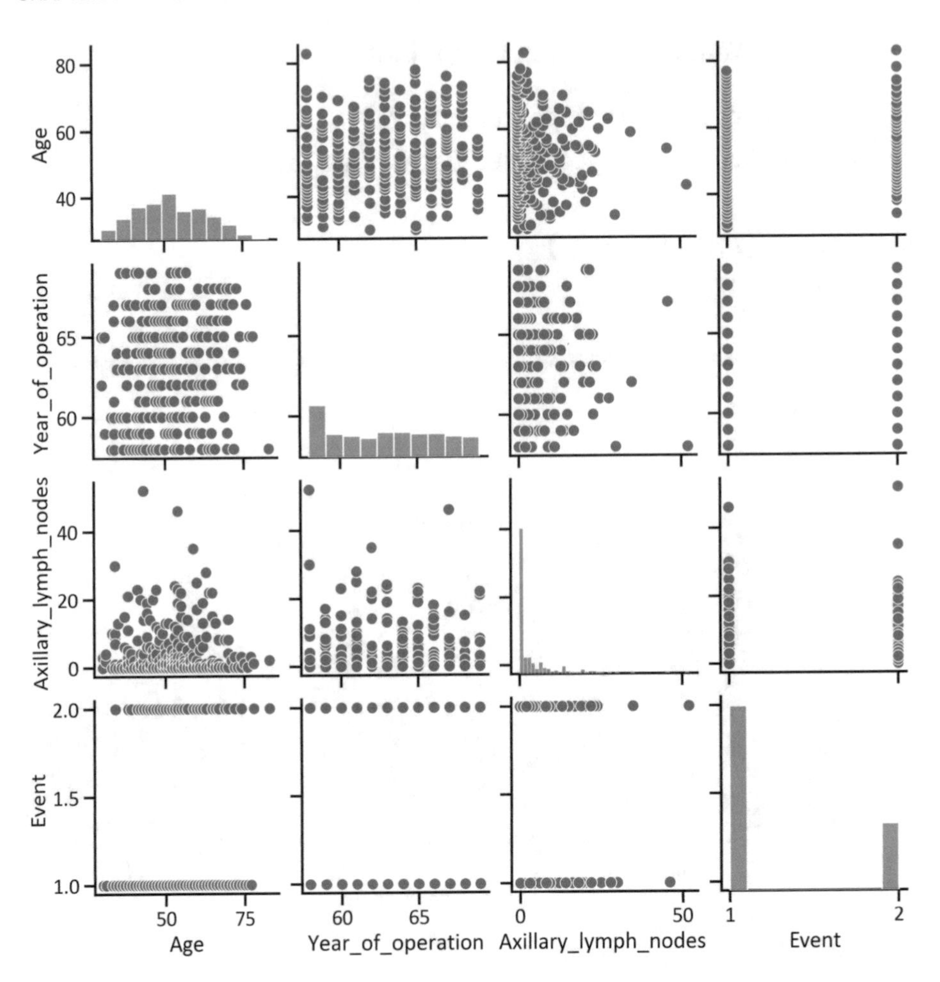

Figure 7-6. *Survival data pair plot*

Figure 7-6 signals values of the features are spread across the scatter. In addition, only Age signals a normal distribution.

Realizing a Correlation Relationship

The Pearson correlation realizes related association between two features, where instances of the features are continuous. For features with classes, you execute other correlation methods (i.e., the Spearman and Kendall correlation methods). Equation 7-2 defines the Pearson correlation method:

$$Corr_{xy} = \frac{Cov_{xy}}{s_x s_y} \quad \text{(Equation 7-2)}$$

where S_x signifies the magnitude at which values of x deviate from $\bar{x}$, S_y signifies the magnitude at which values of y deviates from $\bar{y}$, and Cov_{xy} signifies the magnitude of related changes in x and y. Equation 7-3 defines the covariance method:

$$Cov_{xy} = \frac{\sum_{i=1}^{n}(x_i - \bar{x})(y_i - \bar{y})}{n-1}. \quad \text{(Equation 7-3)}$$

Equation 7-2 generates a correlation coefficient between -1 and 1, where -1 signals an exceedingly negative association, 0 signals the absence of correlation, and 1 signals a remarkably positive association.

Listing 7-6 outlines the Pearson coefficient of the relationship between Age and Year_of_operation (see Table 7-3).

Listing 7-6. Outlining the Pearson Coefficient of the Relationship Between Age and Year of Operation

```
age_yr_op_correlation = survival_censored_data[["Age", "Year_
of_operation"]].corr(method = "pearson")
age_yr_op_correlation
```

Table 7-3. *Pearson Coefficient of the Relationship Between Age and Year of Operation*

	Age	Year_of_operation
Age	1.000000	0.092623
Year_of_operation	0.092623	1.000000

Table 7-3 suggests the presence of a very weak positive correlation relationship between `Year_of_operation` and `Age` (with the Pearson correlation coefficient at 0.0926).

Listing 7-7 outlines the Pearson coefficient of the relationship between `Year_of_operation` and `Event` (see Table 7-4).

Listing 7-7. Outlining the Pearson Coefficient of the Relationship Between Year of Operation and Event

```
yr_op_event_correlation = survival_censored_data[["Year_of_
operation", "Event"]].corr(method = "pearson")
yr_op_event_correlation
```

Table 7-4. *Pearson Coefficient of the Relationship Between Year of Operation and Event*

	Year_of_operation	Event
Year_of_operation	1.000000	-0.004076
Event	-0.004076	1.000000

Table 7-4 suggests the visible presence of an inherently weak negative correlation relationship between `Year_of_operation` and `Event` (with the Pearson correlation coefficient at -0.0040).

Outlining the Survival Table

Survival tables are the core of survival analysis. A survival table keeps track of subjects in a clinical trial. It outlines information like the time at which a subject is included in a study and the time at which a subject abandons the study. Equally, it outlines those at risk.

Listing 7-8 outlines the survival table (see Table 7-5). Start by installing lifelines in your environment: `pip install lifelines`.

Listing 7-8. Cancer Survival Table

```
from lifelines.utils import survival_table_from_events
Time = survival_censored_data["Year_of_operation"]
Event = survival_censored_data["Event"]
survival_table = survival_table_from_events(Time, Event)
survival_table
```

Table 7-5. *Cancer Survival Table*

	removed	observed	censored	entrance	at_risk
event_at					
0.0	0	0	0	305	305
58.0	36	36	0	0	305
59.0	27	27	0	0	269
60.0	28	28	0	0	242
61.0	26	26	0	0	214
62.0	23	23	0	0	188
63.0	30	30	0	0	165
64.0	30	30	0	0	135
65.0	28	28	0	0	105
66.0	28	28	0	0	77
67.0	25	25	0	0	49
68.0	13	13	0	0	24
69.0	11	11	0	0	11

Table 7-5 suggests that in the first year, 305 patients entered the research. By default, all 305 patients were at risk. In 1958, 36 patients left the trial, leaving the trial with 269 the following year. Following that, 28 patients left the study, leaving the trial with 242 patients. By 1969, there were 11 patients at risk.

Carrying Out the Nelson-Aalen Additive Model

Listing 7-9 executes the Nelson-Aalen additive model and outlines the time column (`Year_of_operations`) and event column (`Event`).

Listing 7-9. Carrying Out the Nelson-Aalen Additive Model

```
aalen_additive_method = AalenAdditiveFitter().fit(survival_
censored_data, "Year_of_operation", event_col = "Event")
```

Listing 7-10 outlines the Nelson-Aalen additive model's profile table (see Table 7-6).

Listing 7-10. Outlining the Nelson-Aalen Additive Model's Profile Table

```
aalen_additive_method_summary = aalen_additive_method.print_
summary()
aalen_additive_method_summary
```

Table 7-6. *The Nelson-Aalen Additive Model's Profile Table*

Model	lifelines.AalenAdditiveFitter	
duration col	Year_of_operation	
event col	Event	
number of subjects	305	
number of events observed	305	
time fit was run	2021-09-22 20:45:48 UTC	
	slope(coef)	**se(slope(coef))**
Age	-0.00	0.00
Axillary_lymph_nodes	0.00	0.00
_intercept	0.03	0.12
Concordance	0.47	

Outlining the Nelson-Aalen Additive Model's Confidence Interval

Table 7-6 shows that the coefficient of Age is 0, that of `Axillary_lymph_nodes` is 0, and that of the intercept is 0.03. In addition, the concordance is at 0.47. Listing 7-11 outlines the Nelson-Aalen additive model's confidence interval (Table 7-7).

Listing 7-11. Outlining the Nelson-Aalen Additive Model's Confidence Interval

```
aalen_additive_method_ci = aalen_additive_method.confidence_
intervals_s
aalen_additive_method_ci
```

***Table 7-7.** Outlining the Nelson-Aalen Additive Model's Confidence Interval*

		Age	Axillary_lymph_nodes	_intercept
95% lower-bound	**58.0**	-0.011789	-0.018840	-0.145918
	59.0	-0.018651	-0.021651	-0.004624
	60.0	-0.029200	-0.027498	0.323593
	61.0	-0.031100	-0.032479	0.333276
	62.0	-0.034727	-0.037655	0.354644
	63.0	-0.043889	-0.043974	0.591264
	64.0	-0.052604	-0.054649	0.888015
	65.0	-0.064916	-0.062502	0.977393
	66.0	-0.080685	-0.076952	1.235606
	67.0	-0.108367	-0.111897	1.322455
	68.0	-0.123392	-0.152858	-0.226685
	69.0	-0.287892	-0.262761	-1.590887

(*continued*)

Table 7-7. (*continued*)

		Age	Axillary_lymph_nodes	_intercept
95% upper-bound	**58.0**	0.014053	0.019916	0.258724
	59.0	0.016134	0.022809	0.568260
	60.0	0.015403	0.023818	1.085101
	61.0	0.019462	0.030469	1.196271
	62.0	0.023861	0.041435	1.354825
	63.0	0.026653	0.053084	1.799925
	64.0	0.029656	0.049602	2.307316
	65.0	0.040445	0.066189	2.797615
	66.0	0.051745	0.073717	3.535390
	67.0	0.077604	0.108511	4.569100
	68.0	0.147481	0.129512	4.276818
	69.0	0.311981	0.239415	7.641019

Discerning the Survival Hazard

Listing 7-12 depicts the Nelson-Aalen additive model's survival hazard (see Figure 7-7).

Listing 7-12. Depicting the Nelson-Aalen Additive Model's Survival Hazard

```
aalen_additive_method.plot()
plt.ylabel("Survival hazard")
plt.xlabel("Year of operation")
plt.show()
```

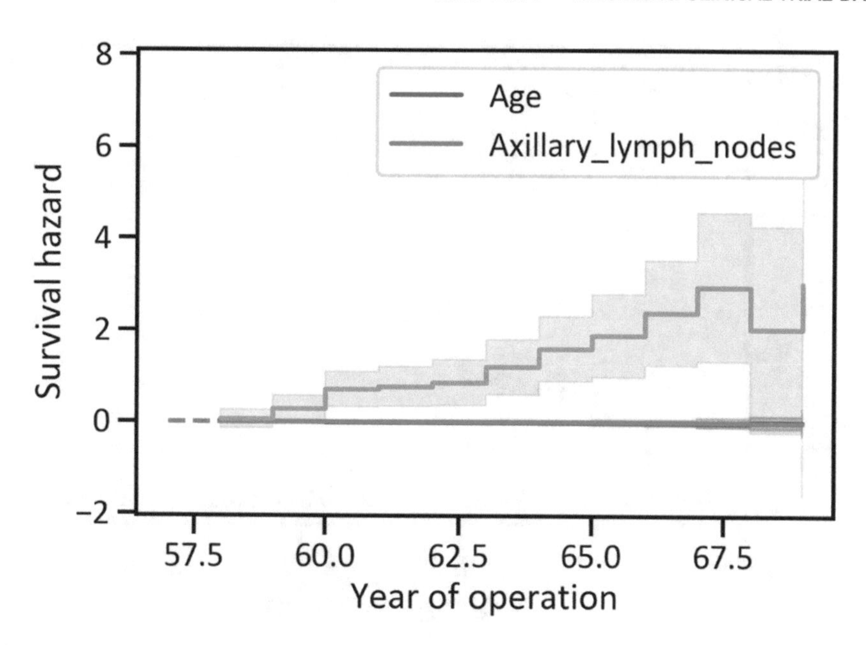

Figure 7-7. *The Nelson-Aalen additive model's survival hazard*

Figure 7-7 signals less activity in relation to Age and Axillary_lymph_nodes. Moreover, the survival hazard escalates from 1958, reaching its peak in 1967, followed by a slight decrease in the subsequent year, and then an increase.

Discerning the Cumulative Survival Hazard

Listing 7-13 depicts the Nelson-Aalen additive model's cumulative survival hazard (see Figure 7-8).

Listing 7-13. Depicting the Nelson-Aalen Additive Model's Cumulative Survival Hazard

```
aalen_additive_method.cumulative_hazards_.plot(lw = 4)
plt.ylabel("Cumulative hazard")
plt.xlabel("Year of operation")
plt.show()
```

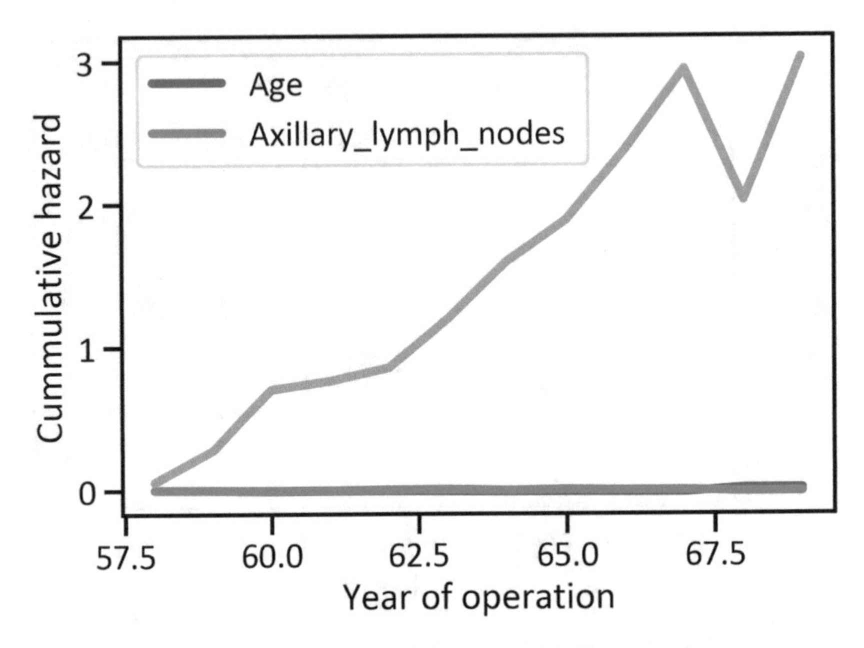

***Figure 7-8.** The Nelson-Aalen additive model's cumulative survival hazard*

Figure 7-8 signals a very strong trend in the cumulative survival hazard from 1958-1966. Moreover, in 1967, the trend lost its momentum, which is marked by a sharp decline in that period. In 1968, a sharp increase erased the decline.

Baseline Survival Hazard

Listing 7-14 depicts the Nelson-Aalen additive model's baseline survival hazard (see Figure 7-9).

Listing 7-14. Depicting the Nelson-Aalen Additive Model's Baseline Survival Hazard

```
aalen_additive_method.cumulative_variance_.plot(lw=4)
plt.ylabel("Cumulative variance")
plt.xlabel("Year of operation")
plt.show()
```

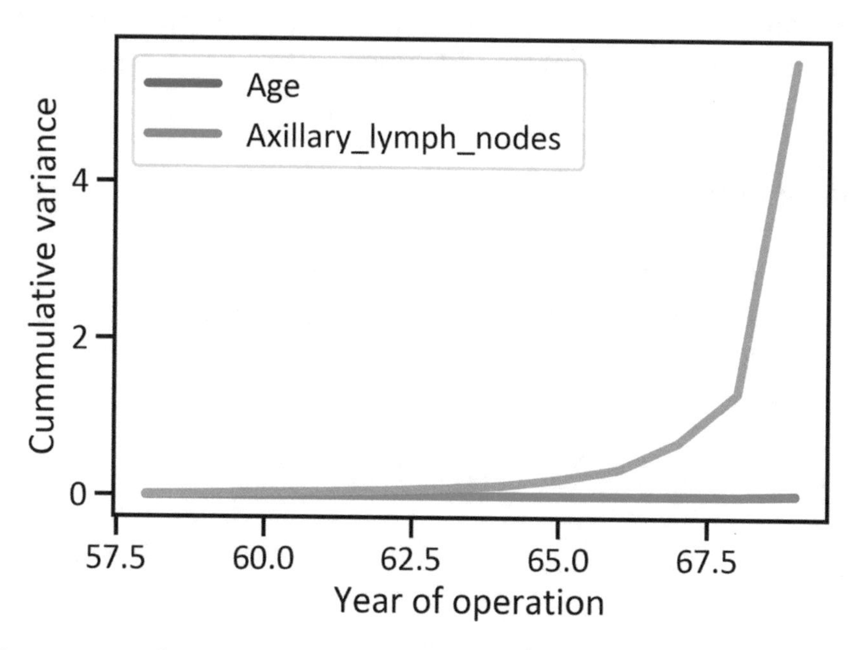

Figure 7-9. *The Nelson-Aalen additive model's cumulative baseline survival hazard*

Figure 7-9 signals what looks like exponential growth in the cumulative variance.

Conclusion

This chapter explained the fundamental principles of carrying out a prevalent survival regressor, recognized as the Nelson-Aalen additive. The method modeled survival censored data to reasonably predict the survival rate of a subject under investigation by carrying out a hazard function, which enables you to predict the likelihood of a patient's survival status across time. There are other methods for survival regression, including Cox Proportional Hazard and Weibull Additive.

Reference

1. University of Chicago's Billings Hospital. (1999). *Haberman's Survival Data*. University of Chicago's Billings Hospital.

CHAPTER 8

Medical Records Categorization

This chapter covers a holistic approach for realizing patterns in medical records by executing a linear discriminant analysis model. You'll learn what medical records are and then you'll learn a technique of cleansing textual data by executing fundamental methods like regularization and `TfidfVectorizer`. Afterward, you'll execute a method to classify the medical specialty and assess the extent to which it segregates classes.

Medical Records

Medical records are historical data relating to the medical history of a patient. General medical practitioners keep records of their patients for reference and decide on reasonable precautions to undertake for a given medical condition. They also keep records for legal reasons such as class action lawsuits where they may present the records in their defense and prosecutors may use them to present a case in the court of law.

T. C. Nokeri, *Artificial Intelligence in Medical Sciences and Psychology*,
https://doi.org/10.1007/978-1-4842-8217-5_8

Context of the Chapter

Traditionally, general medical practitioners maintained their patients' medical history on paper, which was then stored in a secure facility. Today, medical practitioners harness computerized data warehouses.

Medical practitioners may intend to investigate patterns in the medical records of their patients. This chapter introduces an effective technique for executing a classification algorithm to do so. Figure 8-1 depicts the context of the chapter.

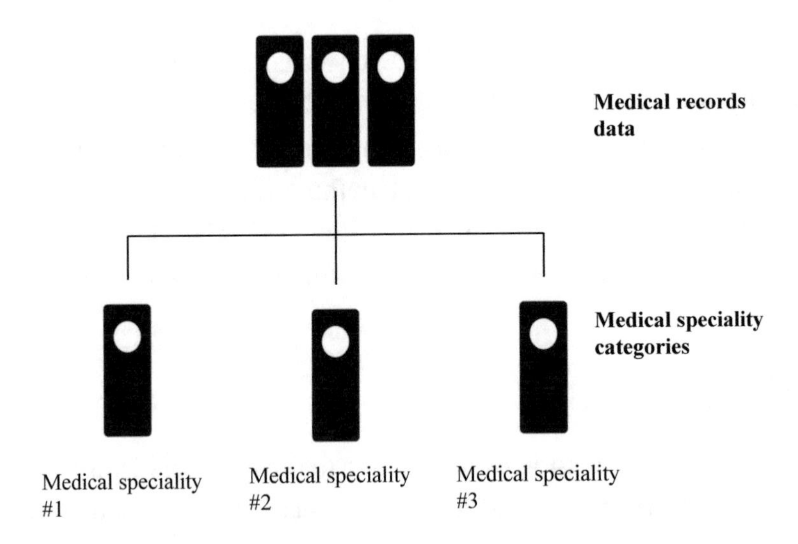

Figure 8-1. *Chapter context*

The exercise may be useful to better patients' medical records and help medical practitioners easily access the records.

You can obtain the dataset from Kaggle[1] (the original dataset comes from mtsamples[2]).

[1] `www.kaggle.com/tboyle10/medicaltranscriptions`

[2] `https://mtsamples.com/`

Categorization with Linear Discriminant Analysis

Linear discriminant analysis makes up a model that bears similarities to linear regression models in the sense that it performs a linear transform. It can be employed for multi-class classification. It carefully ascertains the within-class matrix and between-class matrix while maximizing them (see Equations 8-1, 8-2, and 8-3).

Equation 8-1 defines the within-class matrix:

$$S_{within-class\ matrix} = \sum_{i=1}^{C} \sum_{j=1}^{M_i} \left(y_j - \mu_i\right)\left(y_j - \mu_i\right)^T. \qquad \text{(Equation 8-1)}$$

Equation 8-2 defines the between-class matrix:

$$S_{between-class\ matrix} = \sum_{i=1}^{C} \left(y_j - \mu_i\right)\left(y_j - \mu_i\right)^T. \qquad \text{(Equation 8-2)}$$

Equation 8-3 maximizes the between-class matrix and minimizes the within-class matrix:

$$max = \frac{det\left(S_w\right)}{det\left(S_b\right)}. \qquad \text{(Equation 8-3)}$$

Descriptive Analysis

Listing 8-1 collects the data from a CSV file. Start by installing pandas in your environment: pip install pandas.

Listing 8-1. Data Extraction

```
import pandas as pd
medical_records_data = pd.read_csv(r"filepath\mtsamples.csv")
medical_records_data.drop(['Unnamed: 0'], axis = 1,
inplace = True)
medical_records_data.head()
```

Listing 8-2 depicts the medical specialty count plot (see Figure 8-2). Start by installing Matplotlib in your environment: pip install matplotlib.

Listing 8-2. Medical Specialty Count Plot

```
import matplotlib.pyplot as plt
%matplotlib inline
import seaborn as sns
sns.set("talk","ticks",font_scale=1,font="Calibri")
fig, ax = plt.subplots(figsize=(12, 7))
sns.countplot(medical_records_data["medical_specialty"],
ax = ax)
plt.xticks (rotation=90)
plt.show()
```

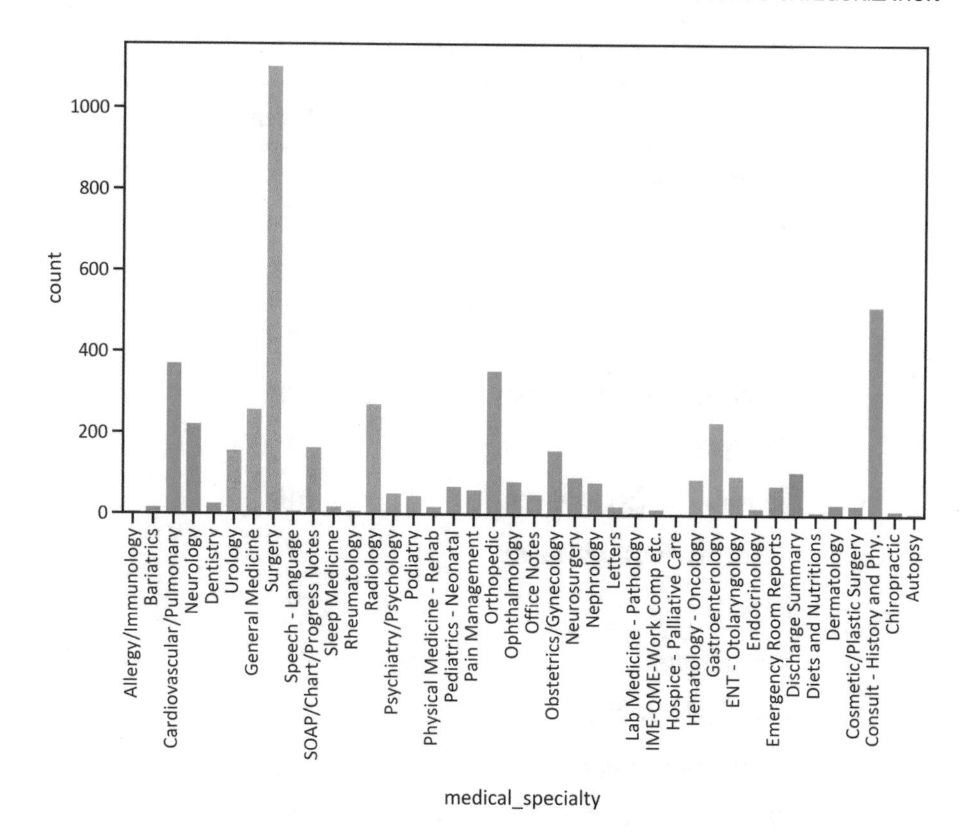

Figure 8-2. *Medical specialty count plot*

Figure 8-2 signals that the `Surgery` medical specialty appears the most in the medical records, followed by `Consult - History and Phy.` and then `Cardiovascular/Pulmonary`. In addition, `Allergy/Immunology` appears the least.

Listing 8-3 counts the number of items in each medical specialty (see Table 8-1).

Listing 8-3. Medical Specialty Count

```
medical_specialty_group = pd.DataFrame(medical_records_data.
groupby("medical_specialty").size())
medical_specialty_group.columns = ["Count"]
medical_specialty_group
```

Table 8-1. *Medical Specialty Count*

	Count
medical_specialty	
Allergy/Immunology	7
Autopsy	8
Bariatrics	18
Cardiovascular/Pulmonary	372
Chiropractic	14
Consult - History and Phy.	516
Cosmetic/Plastic Surgery	27
Dentistry	27
Dermatology	29
Diets and Nutritions	10
Discharge Summary	108
ENT – Otolaryngology	98
Emergency Room Reports	75
Endocrinology	19
Gastroenterology	230
General Medicine	259
Hematology - Oncology	90
Hospice - Palliative Care	6
IME-QME-Work Comp., etc.	16
Lab Medicine – Pathology	8
Letters	23

(*continued*)

Table 8-1. *(continued)*

	Count
Nephrology	81
Neurology	223
Neurosurgery	94
Obstetrics/Gynecology	160
Office Notes	51
Ophthalmology	83
Orthopedic	355
Pain Management	62
Pediatrics - Neonatal	70
Physical Medicine - Rehab	21
Podiatry	47
Psychiatry/Psychology	53
Radiology	273
Rheumatology	10
SOAP/Chart/Progress Notes	166
Sleep Medicine	20
Speech - Language	9
Surgery	1103
Urology	158

Table 8-1 shows that the `Surgery` medical specialty appears 1103 times in the medical records, followed by `Consult - History and Phy.`, which appears 516 times, and `Cardiovascular/Pulmonary`, which appears 372 times. In addition, `Hospice - Palliative` appears the least (6 times).

Preprocessing the Medical Records Data

Listing 8-4 preprocesses the data.

Listing 8-4. Data Preprocessing

```
ind_variables = medical_records_data[["description", "sample_
name", "transcription", "keywords"]].values
labels = medical_records_data["medical_specialty"].values
```

Carrying Out a Regular Expression

Listing 8-5 executes a regular expression to clean the textual data in the medical records. It eliminates special characters, single characters, multiple spaces, and the prefix "b", and then it converts all text to lower cases. Start by installing regex in your environment: pip install regex.

Listing 8-5. Regular Expression

```
import re
clean_ind_variables = []
for sentence in range(len(ind_variables)):
    clean_ind_variable = re.sub(r'\W', ' ',
    str(ind_variables[sentence]))
    clean_ind_variable = re.sub(r'\s+[a-zA-Z]\s+', ' ',
    clean_ind_variable)
    clean_ind_variable = re.sub(r'\^[a-zA-Z]\s+', ' ',
    clean_ind_variable)
    clean_ind_variable = re.sub(r'\s+', ' ', clean_ind_
    variable, flags=re.I)
    clean_ind_variable = re.sub(r'b^\s+', ' ', clean_ind_
    variable)
    clean_ind_variable = clean_ind_variable.lower()
    clean_ind_variables.append(clean_ind_variable)
```

Carrying Out Word Vectorization

Listing 8-6 vectorizes the words in the medical transcript by executing the `TfidfVectorizer()` method and outlining the `max_features` as `2500`, `min_df` as `8`, and `max_df` as `0.8`. The method alters text to draw a `TF-IDF` (frequency-inverse document frequency) features matrix. Start by installing nltk in your environment: pip install nltk.

Listing 8-6. Word Vectorization

```
import nltk
from nltk.corpus import stopwords
from sklearn.feature_extraction.text import TfidfVectorizer
vectorizer = TfidfVectorizer(max_features=2500,
                             min_df=8,
                             max_df=0.7,
                             stop_words=stopwords.
                             words("english"))
clean_ind_variables = vectorizer.fit_transform(clean_ind_
variables).toarray()
```

Listing 8-7 allocates data for training and testing the linear discriminant analysis model. Start by installing scikit-learn in your environment: pip install -U scikit-learn.

Listing 8-7. Allocating Training and Testing Data

```
from sklearn.model_selection import train_test_split
x_train, x_test, y_train, y_test = train_test_split(processed_
features, labels,test_size=0.2,random_state=0)
```

Executing the Linear Discriminant Analysis Model to Classify Patients' Medical Records

Listing 8-8 develops the linear discriminant analysis model and learns the data in the medical records.

Listing 8-8. Executing the Linear Discriminant Analysis Model to Classify Patients' Medical Records

```
from sklearn.discriminant_analysis import
LinearDiscriminantAnalysis
linear_discriminant_model = LinearDiscriminantAnalysis()
linear_discriminant_model.fit(x_train, y_train)
```

Listing 8-9 outlines the linear discriminant analysis model's predictions (see Table 8-2).

Listing 8-9. Computing the Linear Discriminant Analysis Model's Predictions

```
y_hat_linear_discriminant_model = linear_discriminant_model.
predict(x_test)
actual_and_linear_discriminant_model_predictions =
pd.DataFrame({"Actual Medical Specialty": y_test,
                              "Predicted Medical Specialty":
                              y_hat_linear_discriminant_model})
actual_and_linear_discriminant_model_predictions
```

Table 8-2. *Linear Discriminant Analysis Model's Predictions*

	Actual Medical Specialty	Predicted Medical Specialty
0	Consult - History and Phy.	Neurology
1	Radiology	Radiology
2	Surgery	Surgery
3	Surgery	Surgery
4	ENT - Otolaryngology	ENT – Otolaryngology
...	...	...
995	Cardiovascular/Pulmonary	Cardiovascular/Pulmonary
996	Orthopedic	General Medicine
997	Surgery	Autopsy
998	Surgery	Bariatrics
999	SOAP/Chart/Progress Notes	SOAP/Chart/Progress Notes

Considering the Linear Discriminant Analysis Model's Performance

Listing 8-10 outlines the artificial neural network's classification report, which holds the accuracy score, precision score, recall, f-1 score, and support (see Table 8-3).

Listing 8-10. Linear Discriminant Analysis Classification Report

```
from sklearn import metrics
linear_discriminant_model_report = pd.DataFrame(metrics.
classification_report(y_test,
             y_hat_linear_discriminant_model,
             output_dict=True)).transpose()
linear_discriminant_model_report
```

Table 8-3. Linear Discriminant Analysis Classification Report

	Precision	Recall	F-1 score	Support
Allergy/Immunology	0.200000	1.000000	0.333333	1.000
Autopsy	0.018868	1.000000	0.037037	1.000
Bariatrics	0.000000	0.000000	0.000000	0.000
Cardiovascular/ Pulmonary	0.571429	0.555556	0.563380	72.000
Chiropractic	0.000000	0.000000	0.000000	1.000
Consult - History and Phy.	0.421053	0.273504	0.331606	117.000
Cosmetic/Plastic Surgery	0.454545	0.833333	0.588235	6.000
Dentistry	0.000000	0.000000	0.000000	1.000
Dermatology	0.000000	0.000000	0.000000	6.000
Diets and Nutritions	0.000000	0.000000	0.000000	0.000
Discharge Summary	0.705882	0.545455	0.615385	22.000
ENT - Otolaryngology	0.615385	0.470588	0.533333	17.000
Emergency Room Reports	0.357143	0.357143	0.357143	14.000
Endocrinology	0.000000	0.000000	0.000000	3.000
Gastroenterology	0.790698	0.693878	0.739130	49.000
General Medicine	0.513514	0.345455	0.413043	55.000
Hematology - Oncology	0.333333	0.307692	0.320000	13.000
Hospice - Palliative Care	0.000000	0.000000	0.000000	0.000
IME-QME-Work Comp., etc.	0.000000	0.000000	0.000000	0.000
Lab Medicine - Pathology	0.200000	1.000000	0.333333	1.000
Letters	0.000000	0.000000	0.000000	4.000

(continued)

Table 8-3. *(continued)*

	Precision	Recall	F-1 score	Support
Nephrology	0.096774	0.272727	0.142857	11.000
Neurology	0.733333	0.622642	0.673469	53.000
Neurosurgery	0.750000	0.857143	0.800000	21.000
Obstetrics/Gynecology	0.807692	0.700000	0.750000	30.000
Office Notes	0.714286	0.769231	0.740741	13.000
Ophthalmology	0.750000	0.818182	0.782609	11.000
Orthopedic	0.709091	0.609375	0.655462	64.000
Pain Management	0.391304	0.562500	0.461538	16.000
Pediatrics - Neonatal	0.000000	0.000000	0.000000	14.000
Physical Medicine – Rehab	0.066667	0.250000	0.105263	4.000
Podiatry	0.000000	0.000000	0.000000	8.000
Psychiatry/Psychology	0.000000	0.000000	0.000000	12.000
Radiology	0.904762	0.666667	0.767677	57.000
Rheumatology	0.000000	0.000000	0.000000	2.000
SOAP/Chart/Progress Notes	0.793103	0.560976	0.657143	41.000
Sleep Medicine	0.046154	0.600000	0.085714	5.000
Speech - Language	0.000000	0.000000	0.000000	0.000
Surgery	0.809859	0.527523	0.638889	218.000
Urology	0.727273	0.648649	0.685714	37.000
Accuracy	0.508000	0.508000	0.508000	0.508
Macro avg	0.337054	0.396205	0.327801	1000.000
Weighted avg	0.632999	0.508000	0.554631	1000.000

Table 8-3 shows that the linear discriminant analysis model is more precise when predicting that records fall under the `Radiology` medical specialty (with the precision rate at 90%). It is less precise when predicting that records fall under the `Endocrinology`, `Hospice - Palliative Care`, `Letters`, `Pediatrics - Neonatal`, `Psychiatry/Psychology`, and `Speech-Language` (with the precision rate at 0%). In addition, the model is 50% accurate when distinguishing the medical specialty classes.

Conclusion

In this chapter, you carried out a multi-class classification model recognized as linear discriminant analysis. Since medical records are composed mostly of textual data, you took a holistic approach to cleansing text by executing the regular expression and the tokenization method. You also found a way of ascertaining the extent to which the discriminant model reliably distinguishes medical specialty classes by employing a classification report.

CHAPTER 9

A Case for Psychology: Factoring and Clustering Personality Dimensions

This chapter covers how to analyze the underlying patterns in human behavior by carrying out exploratory factor analysis and cluster analysis. To begin, you'll learn about the big five personality dimensions. Following that, you'll explore an approach for collecting data by retaining a Likert scale and measuring the reliability of the scale with Cronbach's reliability testing strategy. Subsequently, you'll perform factor analysis beginning with estimating Bartlett Sphericity statistics and then the Kaiser-Meyer-Olkin statistic. Following that, you'll rotate the eigenvalues by carrying out the varimax rotation method and estimate the proportional variances and cumulative variances. In addition, you'll execute the K-Means method to observe clusters in the data beginning with standardizing the data and carrying out principal component analysis.

T. C. Nokeri, *Artificial Intelligence in Medical Sciences and Psychology*,
https://doi.org/10.1007/978-1-4842-8217-5_9

Personality Dimensions

A person's personality is a combination of traits that make that person unique and reflect consistent patterns in behavior. The most standard way for underpinning human behavior involves the use of the big five personality dimensions test, which groups several traits (see Table 9-1).

Table 9-1. *Big Five Personality Dimensions*

Personality trait	Description
Extraversion	Signifies traits that reflect the inclination towards socializing and confidence. Contrasting traits are introversion, constituting self-reservedness, seriousness, and quietness, among others.
Neuroticism	Signifies traits that reflect the inclination towards anxiety, tenseness, low self-esteem, and irrationality, among others. The contrasting traits is stability.
Conscientious	Signifies traits that reflect the inclination towards obedience, impulse control, and integrity, among other traits. The contrasting trait is expediency.
Openness to experience	Signifies traits that reflect the inclination towards new ideas, intellectual pursuits, and diverse interests, among others. The opposite is closedness to experience.
Agreeableness	Signifies traits that reflect the inclination towards trusting, empathy, and cooperativeness, among others. Contrasting traits are disagreeability or hostility.

Learn more about the big five personality dimensions from the official Open Psychometrics website[1].

[1] https://ipip.ori.org/new_ipip-50-item-scale.htm

Questionnaires

A questionnaire is a set of questions that are structured in such a way that they capture the underlying behavior of a subject, so we can examine patterns in the behavior of a set of subjects. A basic questionnaire comprises two sections: biogeographical factors and the Likert scale.

Likert Scale

The most convenient approach for studying human behavior involves administering questionnaires to the subject (referred to as self-administered questionnaires). To capture variability in data among a set of subjects, we standardize the questionnaire using some scale (i.e., the Likert scale), entailing that all subjects receive the same questionnaire.

Table 9-2 exhibits the structure of a Likert scale.

Table 9-2. *Structure of a Likert Scale*

	Strongly disagree	Disagree	Neither disagree nor agree	Agree	Strongly agree
Question item	-	-	-	-	-

Figure 9-1 exhibits the structure of a Likert scale where we code `"strongly disagree"` as `1`, `"disagree"` as `2`, `"neither disagree nor agree"` as `3`, `"agree"` as `4`, and `"strongly agree"` as `5`.

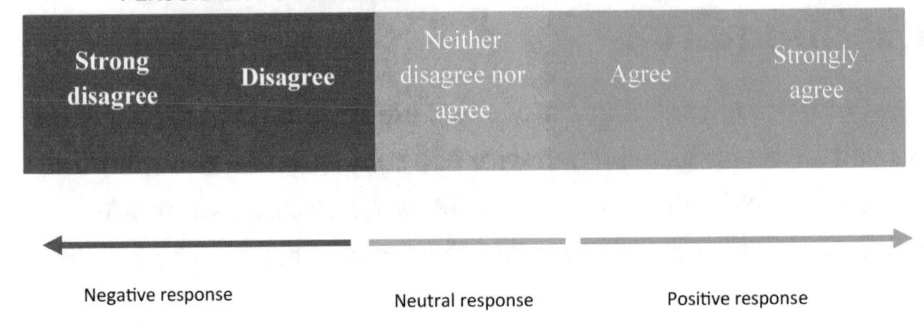

Figure 9-1. *Negative and positive responses*

This chapter applies a factor model and a K-Means cluster to model the big five personality dataset provided by Open Psychometrics[2]. You may download the dataset from Kaggle[3].

Listing 9-1 collects the data. Start by installing pandas in your environment: `pip install pandas`.

Listing 9-1. Collecting the Data

```
import pandas as pd
big_five_data = pd.read_csv(r"filepath\big_five_data\
data-final.csv", sep = "\t")
big_five_data
```

Listing 9-2 preprocesses the data.

Listing 9-2. Preprocessing the Data

```
big_five_data.drop(big_five_data.columns[50:110], axis = 1,
inplace = True)
big_five_data = big_five_data[(big_five_data > 0).all(axis=1)]
```

[2] https://openpsychometrics.org/tests/IPIP-BFFM/
[3] www.kaggle.com/tunguz/big-five-personality-test

Scale Reliability

There are innumerable scale reliability testing strategies. This chapter only acquaints you with two prevalent reliability testing strategies: Spearman-Brown and Cronbach's reliability testing.

Spearman-Brown Reliability Testing Strategy

Equation 9-1 defines the Spearman-Brown reliability test:

$$r_{tt} = \frac{2r_h}{1+r_h} \qquad \text{(Equation 9-1)}$$

where r_{tt} constitutes the reliability and r_h constitutes the correlation between the two halves of the test.

Carrying Out the Cronbach's Reliability Testing Strategy

This section shows that Cronbach's coefficient alpha technique captures the overall reliability. Internal consistency represents the degree to which the responses are consistent with the items within a specific measure.

Equation 9-2 defines Cronbach's reliability test:

$$\rho_T = \frac{k^2 - \sum_{i=1}^{n} \sigma_i}{\sigma_i^2} \qquad \text{(Equation 9-2)}$$

where ρ_T constitutes the reliability test, k constitutes the number of items in the Likert scale, σ_{ij} constitutes how X_i and X_j, and σ_X^2 constitutes the variance and covariance among items.

Listing 9-3 discerns the Cronbach's coefficient alpha. Start by installing pingouin in your environment: `pip install pingouin.`

Listing 9-3. Discerning the Cronbach's Coefficient Alpha

```
import pingouin as pg
pg.cronbach_alpha(data = big_five_data)
```

```
(0.5199338748916366, array([0.518, 0.521]))
```

The findings above suggest that Cronbach's coefficient alpha is 0.5199, which suggests the measurement is not exemplary. Moreover, there is mediocre consistency with the items in the measurement scale.

Carrying Out the Factor Model

Factor analysis is an alternative dimension reduction technique. It equally considers the variability in factors. Although it shares considerable similarities with principal component analysis, before carrying out it, there are a series of steps that you ought to take (i.e., recognizing the sample adequacy). Figure 9-2 exhibits how factor analysis functions.

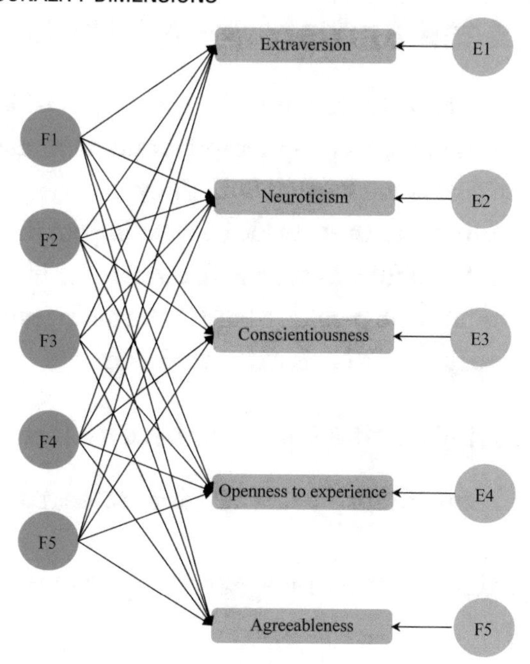

Figure 9-2. *Factor model*

Figure 9-2 constitutes features (extraversion, neuroticism, conscientiousness, openness to experience, and agreeableness) grounded on factors (`F1`, `F2`, `F3`, and `F4`). Equations 9-3, 9-4, 9-5, 9-6, and 9-7 express Figure 9-2:

$$Extraversion_1 = \beta_{10} + \beta_{11}F_1 + \beta_{12}F_2 + \varepsilon_i \quad \text{(Equation 9-3)}$$

$$Neuroticism_2 = \beta_{20} + \beta_{21}F_2 + \beta_{12}F_2 + \varepsilon_i \quad \text{(Equation 9-4)}$$

$$Conscientious_3 = \beta_{30} + \beta_{31}F_3 + \beta_{13}F_3 + \varepsilon_i \quad \text{(Equation 9-5)}$$

$$Openness\ to\ experience_4 = \beta_{40} + \beta_{41}F_4 + \beta_{14}F_4 + \varepsilon_i \quad \text{(Equation 9-6)}$$

$$Agreeableness_5 = \beta_{50} + \beta_{51}F_1 + \beta_{52}F_5 + \varepsilon_i \quad \text{(Equation 9-7)}$$

Carrying Out the Bartlett Sphericity Test

Prior to performing factor analysis, initially you must decide whether the data is suitable. The most prevalent method for unanimously ratifying the appropriateness of data is the Bartlett Sphericity.

Listing 9-4 puts together the p-value by carrying out the `calculate_bartlett_sphericity` method in the `factor_analyzer` library (see Table 9-3 , which returns the p-value). Start by installing factor_analyzer in your environment: `pip install factor_analyzer`.

Listing 9-4. Carrying Out the Bartlett Sphericity Test

```
from factor_analyzer.factor_analyzer import calculate_bartlett_
sphericity
chi_square, p_values = calculate_bartlett_sphericity
(big_five_data)
chi_square = pd.DataFrame(pd.Series(chi_square),columns=
["Chi squared"])
p_values = pd.DataFrame(pd.Series(p_values),columns = ["P value"])
barlett_sphericity = pd.concat([chi_square, p_values],axis=1)
barlett_sphericity = barlett_sphericity.transpose()
barlett_sphericity.columns = ["Bartlett sphericity"]
barlett_sphericity
```

Table 9-3. *Bartlett Sphericity Statistics*

	Bartlett sphericity
Chi squared	1.717619e+07
P value	0.000000e+00

Carrying Out the Kaiser-Meyer-Olkin Test

It is a traditional practice, especially in academic research, to decide whether data is worth interrogating. A prevalent approach for doing so involves carrying out the Kaiser-Meyer-Olkin statistic.

Listing 9-5 puts together the p-value by carrying out the `calculate_kmo()` method in the `factor_analyzer` library (see Table 9-4, which returns the p-value).

Listing 9-5. Carrying Out the KMO Test

```
from factor_analyzer.factor_analyzer import calculate_kmo
kaiser_meyer_olkin_variance, kaiser_meyer_olkin_test =
calculate_kmo(big_five_data)
kaiser_meyer_olkin_test = pd.DataFrame(pd.Series(kaiser_meyer_
olkin_test))
kaiser_meyer_olkin_test.columns = ["KMO Test"]
kaiser_meyer_olkin_test.index = ["P value"]
kaiser_meyer_olkin_test
```

Table 9-4. *KMO test statistics*

	KMO test
P value	0.908983

Table 9-4 shows that the p-value > 0.5, so the data is appropriate. After deciding on the appropriateness of the data, the subsequent step involves deciding on a number of factors.

Discerning K with a Scree Plot

The most prevalent way for wisely deciding on the number of factors involves constructing a scree plot, which constitutes factors in the x-axis and Eigen loadings or eigenvalues (vectors from a linear transformation). To determine the number of factors, detect an area where the curve brings into being an extreme bend.

Listing 9-6 puts together a scree plot (see Figure 9-3). Start by installing Matplotlib in your environment: `pip install matplotlib`.

Listing 9-6. Putting Together a Scree Plot

```
import matplotlib.pyplot as plt
%matplotlib inline
from factor_analyzer import FactorAnalyzer
factor_analys_model = FactorAnalyzer(rotation = None,
impute = "drop", n_factors = big_five_data.shape[1])
factor_analys_model.fit(big_five_data)
eigenvalues,_ = factor_analys_model.get_eigenvalues()
plt.scatter(range(1, big_five_data.shape[1]+1), eigenvalues,
s = 200, color = "blue")
plt.plot(range(1, big_five_data.shape[1]+1), eigenvalues,
lw = 4, color = "orange")
plt.xlabel("n factors")
plt.ylabel("Eigenvalues")
plt.show()
```

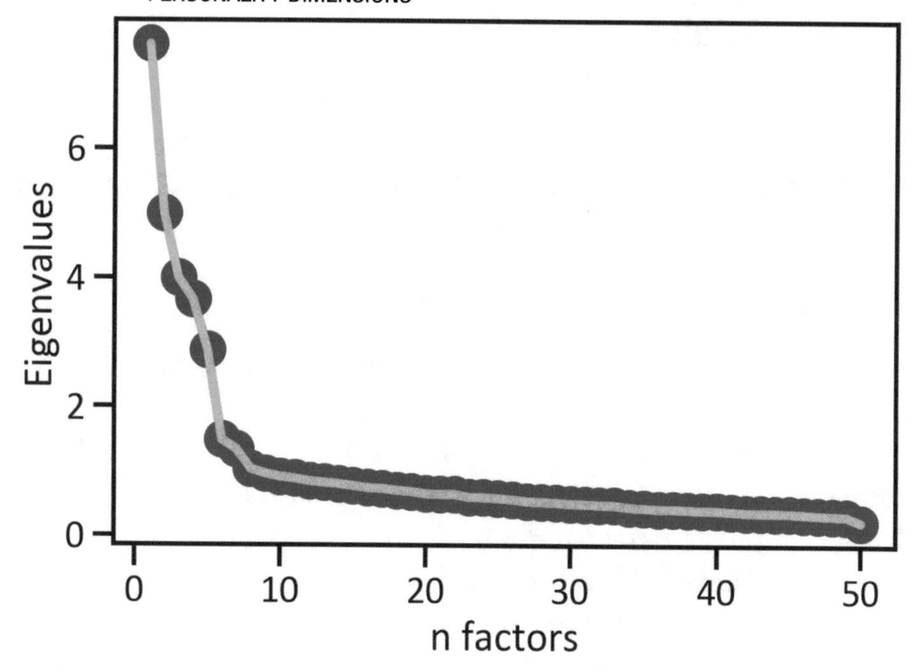

Figure 9-3. *Scree plot*

Figure 9-3 shows that the curve begins to bend extremely from the second factor. As a result, the factor analysis model will be fitted containing two factors.

Carrying Out Eigenvalue Rotation

After fitting the factor analysis model, the subsequent step typically involves rotating the eigenvalue. There are innumerable rotation techniques in the `factor_analyzer` library (i.e., `promax`, `oblique`, `oblimin`, and `oblimax`, among others). The next section shows how to satisfactorily perform the varimax rotation.

Varimax Rotation

A varimax rotation serves to streamline factor loadings.

Listing 9-7 puts together factor loadings by carrying out the FactorAnalyzer() method and then it specifies the number of factors and the rotation method (see Table 9-5).

Listing 9-7. Varimax Transformation

```
factor_analysis_model = FactorAnalyzer(n_factors = 5,
rotation = "varimax")
factor_analysis_model.fit(big_five_data)
rotation = pd.DataFrame(factor_analysis_model.loadings_,
index = big_five_data.columns)
rotation.columns = ["Factor 1",
                    "Factor 2",
                    "Factor 3",
                    "Factor 4",
                    "Factor 5"]
rotation
```

Table 9-5. *Rotation*

	Factor 1	Factor 2	Factor 3	Factor 4	Factor 5
EXT1	0.695725	-0.041532	0.071403	-0.009142	0.017345
EXT2	-0.701616	-0.006997	-0.132951	0.020675	-0.039147
EXT3	0.642162	-0.258014	0.255137	0.095699	-0.025813
EXT4	-0.738570	0.115534	-0.050113	-0.027933	-0.005537
EXT5	0.720532	-0.075416	0.210091	0.082719	0.069474
EXT6	-0.545367	0.059033	-0.139545	-0.026986	-0.260178

(*continued*)

Table 9-5. *(continued)*

	Factor 1	Factor 2	Factor 3	Factor 4	Factor 5
EXT7	0.716289	-0.087368	0.152714	0.025976	0.016274
EXT8	-0.595159	0.031824	0.040168	0.069637	-0.048275
EXT9	0.639384	-0.051900	-0.029303	-0.045866	0.126483
EXT10	-0.676138	0.161479	-0.063951	-0.030044	-0.013397
EST1	-0.116700	0.710429	0.106147	-0.004203	-0.081596
EST2	0.102799	-0.549918	-0.003257	-0.044860	0.030262
EST3	-0.143538	0.618867	0.187814	0.041708	-0.004819
EST4	0.138627	-0.370749	-0.035189	0.098716	-0.076226
EST5	-0.042177	0.505833	-0.007284	-0.079965	-0.113196
EST6	-0.033118	0.738338	0.020280	-0.067361	-0.094172
EST7	0.018081	0.713498	-0.029624	-0.159438	-0.016969
EST8	0.000966	0.737296	-0.044973	-0.167907	-0.021169
EST9	-0.020265	0.692480	-0.175702	-0.039147	-0.051407
EST10	-0.247948	0.610271	-0.023933	-0.188655	0.083560
AGR1	-0.017035	0.025051	-0.489886	-0.027460	-0.083585
AGR2	0.343977	-0.043422	0.548944	-0.001409	0.097968
AGR3	0.118045	0.212494	-0.404620	-0.184223	0.052855
AGR4	0.032340	0.080825	0.791517	0.032914	0.028774
AGR5	-0.133649	-0.004941	-0.659763	-0.002029	-0.031660
AGR6	-0.011192	0.169866	0.600732	0.019605	-0.054179
AGR7	-0.296800	0.074615	-0.633004	-0.010930	-0.056301
AGR8	0.139934	-0.002825	0.566264	0.093890	0.028801

(continued)

Table 9-5. *(continued)*

	Factor 1	Factor 2	Factor 3	Factor 4	Factor 5
AGR9	0.092254	0.129425	0.695292	0.048768	0.060667
AGR10	0.311460	-0.123720	0.397421	0.120033	0.091863
CSN1	0.030204	-0.086851	0.015461	0.642191	0.079651
CSN2	0.056812	0.102458	0.037685	-0.563525	0.117736
CSN3	-0.034077	0.029644	0.084773	0.402538	0.240358
CSN4	-0.040437	0.348466	-0.032655	-0.576761	0.000837
CSN5	0.072870	-0.077612	0.046832	0.622983	-0.085307
CSN6	0.015893	0.162743	-0.001254	-0.612138	0.052051
CSN7	-0.047800	0.089839	0.027289	0.559690	0.022201
CSN8	-0.040542	0.206362	-0.131315	-0.482138	-0.051629
CSN9	0.054589	0.028122	0.102645	0.618940	-0.063249
CSN10	0.031027	-0.010591	0.044011	0.453525	0.239131
OPN1	0.039073	-0.028388	-0.043922	0.047399	0.589616
OPN2	-0.005520	0.186251	-0.020297	-0.003673	-0.581455
OPN3	0.040570	0.114390	0.091881	-0.089634	0.542348
OPN4	0.022583	0.088703	-0.108734	0.069783	-0.521393
OPN5	0.216395	-0.076759	-0.027234	0.136569	0.574784
OPN6	-0.071497	0.031979	-0.099417	0.029790	-0.503171
OPN7	0.071966	-0.147081	-0.018945	0.173371	0.471799
OPN8	0.029606	0.061729	-0.115983	-0.022972	0.552540
OPN9	-0.130261	0.132776	0.172660	0.043014	0.388514
OPN10	0.185348	-0.010013	0.036221	0.017484	0.663781

Discerning Proportional Variance and Cumulative Variances

Listing 9-8 puts together factor loadings by carrying out the `FactorAnalyzer()` method and then it specifies the number of factors and the rotation method (see Table 9-6).

Listing 9-8. Computing Proportional Variance and Cumulative Variances

```
factor_analysis_model_variance = pd.DataFrame(factor_analysis_
model.get_factor_variance(), index=["Variance",
                                     "Proportional Variance",
                                     "Cumulative Variance"])
factor_analysis_model_variance.columns = ["Factor 1",
                                          "Factor 2",
                                          "Factor 3",
                                          "Factor 4",
                                          "Factor 5"]
factor_analysis_model_variance
```

Table 9-6. *Proportional Variance and Cumulative Variances*

	Communalities
EXT1	0.491242
EXT2	0.511950
EXT3	0.553863
EXT4	0.562156
EXT5	0.580662

(*continued*)

Table 9-6. *(continued)*

	Communalities
EXT6	0.388804
EXT7	0.544965
EXT8	0.364021
EXT9	0.430465
EXT10	0.488410
EST1	0.536271
EST2	0.315916
EST3	0.440637
EST4	0.173465
EST5	0.276907
EST6	0.560057
EST7	0.535992
EST8	0.574270
EST9	0.514986
EST10	0.477055
AGR1	0.248647
AGR2	0.431145
AGR3	0.259537
AGR4	0.635988
AGR5	0.454181
AGR6	0.393179
AGR7	0.497641

(continued)

Table 9-6. (*continued*)

	Communalities
AGR8	0.349889
AGR9	0.514751
AGR10	0.293104
CSN1	0.427448
CSN2	0.346568
CSN3	0.229035
CSN4	0.456783
CSN5	0.408912
CSN6	0.404162
CSN7	0.324846
CSN8	0.296595
CSN9	0.401394
CSN10	0.265880
OPN1	0.354155
OPN2	0.373235
OPN3	0.325348
OPN4	0.296921
OPN5	0.402488
OPN6	0.270087
OPN7	0.279822
OPN8	0.323967
OPN9	0.217202
OPN10	0.476676

Carrying Out Cluster Analysis

This section carries out the K-Means model to discern distinct clusters in the big five personality dimensions dataset. Initially, you standardize the data, and then you perform principal component analysis. The model computes the considerable distance between instances of the data to realize similar instances, and then it figures out the centers of the cluster.

Equation 9-8 defines K-Means:

$$J = \sum_{j=1}^{k}\sum_{i=1}^{n}\left|x_i^j - C_j\right| \quad \text{(Equation 9-8)}$$

where J signifies the objective function, k signifies the number of clusters, n signifies the number of cases, x signifies C_j, and C_j signifies the centroids of j.

The K-Means models show data reduced with principal component analysis.

Carrying Out Principal Component Analysis

The principal component analysis technique enables you to investigate inconsistencies in the data. It considers the extent to which variability in the data describes the model. When carrying out this technique, you focus on components equally recognized as latent variables. A component represents a variable that derives from a quantitative model in place of observation.

Figure 9-4 demonstrates an abstract model, where each component represents a variable of interest.

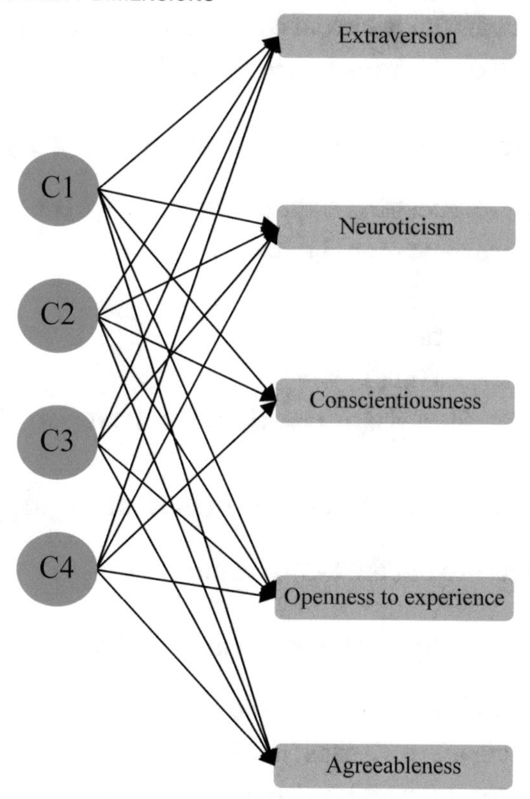

Figure 9-4. *Principal components*

Figure 9-4 constitutes features (`extraversion`, `neuroticism`, `conscientiousness`, `openness to experience`, and `agreeableness`) grounded on components (`C1`, `C2`, `C3`, and `C4`). A component that remarkably considers the variability constitutes the first component, the subsequent component constitutes the second component, and so forth.

Equation 9-9 defines the PCA model:

$$\sum_{i=1}^{n} \frac{1}{m}(x_t - \overline{x})(x_i - \overline{x})^T \qquad \text{(Equation 9-9)}$$

where $\sum_{i=1}^{n} x_i$ is the eigenvalue.

Listing 9-9 carries out principal component analysis. Start by installing scikit-learn in your environment: `pip install scikit-learn`.

Listing 9-9. Carrying Out Principal Component Analysis

```
from sklearn.preprocessing import StandardScaler
from sklearn.decomposition import PCA
pca = PCA(n_components = 5)
pca.fit(big_five_data)
scaler = StandardScaler()
standardized_big_five_data = scaler.fit_transform(big_
five_data)
pca_big_five_data = pca.transform(standardized_big_five_data)
```

Listing 9-10 carries out the K-Means model.

Listing 9-10. Carrying Out the K-Means Model

```
from sklearn.cluster import KMeans
big_five_data_kmeans = KMeans(n_clusters = 5)
big_five_data_kmeans_model = kmeans.fit(pca_big_five_data)
big_five_data_kmeans_model
```

Returning K-Means Labels

Listing 9-11 returns K-Means labels (see Table 9-7).

Listing 9-11. Returning K-Means Labels

```
big_five_data_kmeans_labels = pd.DataFrame(big_five_data_
kmeans_model.labels_, columns = ["Clusters"])
big_five_data_kmeans_labels
```

Table 9-7. K-Means Labels

	Clusters
0	2
1	3
2	2
3	0
4	2
...	...
874429	3
874430	0
874431	1
874432	0
874433	1

Discerning K-Means Cluster Centers

Listing 9-12 discerns K-Means cluster centers (Table 9-8)

Listing 9-12. Discerning K-Means Cluster Centers

```
big_five_data_kmeans_centers = big_five_data_kmeans_model.
cluster_centers_big_five_data_kmeans_centroids =
pd.DataFrame(big_five_data_kmeans_centers).transpose()
big_five_data_kmeans_centroids.columns = ["Cluster 1",
"Cluster 2", "Cluster 3", "Cluster 4", "Cluster 5"]
big_five_data_kmeans_centroids
```

Table 9-8. *K-Means Cluster Centers*

	Cluster 1	Cluster 2	Cluster 3	Cluster 4	Cluster 5
0	1.070171	-2.365766	-4.186860	0.702053	0.049696
1	6.917444	5.127327	8.496112	8.223140	10.037378
2	5.106830	6.945864	5.599746	6.354891	8.202415
3	1.625516	2.770158	2.578343	4.617922	2.585453
4	-5.435118	-5.658154	-5.808755	-7.238314	-4.330565

Listing 9-13 displaying K-Means labels together with the cluster centers (see Figure 9-5).

Listing 9-13. Displaying K-Means Labels

```
fig, ax = plt.subplots()
plt.scatter(pca_big_five_data[:,0], pca_big_five_data[:,1],
c=big_five_data_kmeans_model.labels_, cmap = "coolwarm",s = 15)
plt.scatter(big_five_data_kmeans_centers[:,0], big_five_data_
kmeans_centers[:,1], color="black")
plt.xlabel("y")
plt.show()
```

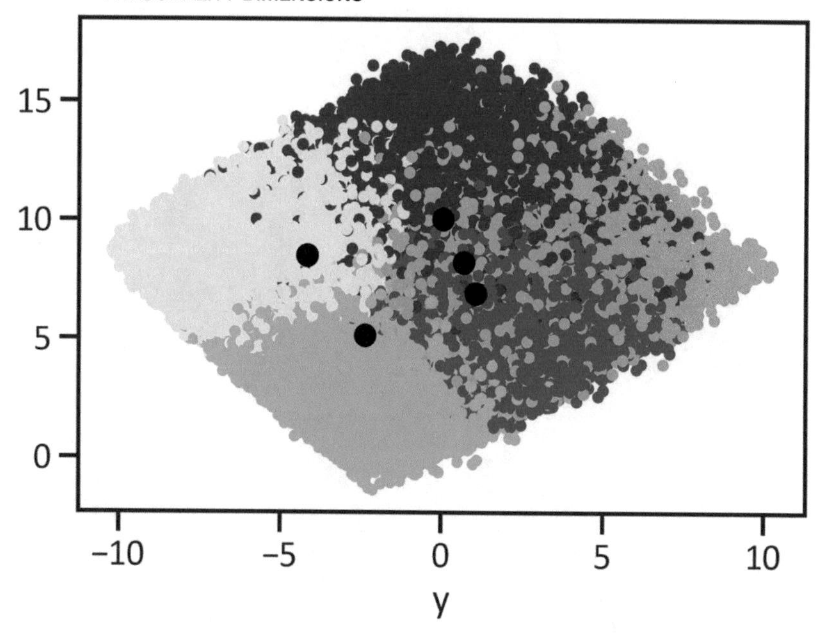

Figure 9-5. *K-Means results*

Conclusion

This chapter concluded the book by presenting a holistic approach for analyzing distinct personalities by carrying out a factor analysis and a cluster analysis. You fitted the cluster model with the reduced data deriving from the principal component analysis method.

I hope you find this book useful and I hope it helps you comprehend ways of tackling various problems across several fields in the medical sciences. Please spread the word and share the book with your counterparts.

Thank you for your time.

Index

T. C. Nokeri, *Artificial Intelligence in Medical Sciences and Psychology*,
https://doi.org/10.1007/978-1-4842-8217-5

D, E

F

G

H, I, J

K

X, Y, Z

Printed in the United States
by Baker & Taylor Publisher Services